Advanced Structured Materials

Volume 229

Series Editors

Andreas Öchsner, Faculty of Mechanical and Systems Engineering, Esslingen University of Applied Sciences, Esslingen, Germany

Lucas F. M. da Silva, Department of Mechanical Engineering, Faculty of Engineering, University of Porto, Porto, Portugal

Holm Altenbach, Faculty of Mechanical Engineering, Otto von Guericke University Magdeburg, Magdeburg, Germany

Common engineering materials are reaching their limits in many applications, and new developments are required to meet the increasing demands on engineering materials. The performance of materials can be improved by combining different materials to achieve better properties than with a single constituent, or by shaping the material or constituents into a specific structure. The interaction between material and structure can occur at different length scales, such as the micro, meso, or macro scale, and offers potential applications in very different fields.

This book series addresses the fundamental relationships between materials and their structure on overall properties (e.g., mechanical, thermal, chemical, electrical, or magnetic properties, etc.). Experimental data and procedures are presented, as well as methods for modeling structures and materials using numerical and analytical approaches. In addition, the series shows how these materials engineering and design processes are implemented and how new technologies can be used to optimize materials and processes.

Advanced Structured Materials is indexed in Google Scholar and Scopus.

Tajul Adli Abdul Razak · Ahmad Kamal Ismail ·
Andreas Öchsner
Editors

Progress in Engineering Technology VI

Springer

Editors
Tajul Adli Abdul Razak
Malaysian Spanish Institute
University of Kuala Lumpur
Kulim, Kedah, Malaysia

Ahmad Kamal Ismail
Malaysian Spanish Institute
University of Kuala Lumpur
Kulim, Kedah, Malaysia

Andreas Öchsner
Faculty of Mechanical and Systems Engineering
Esslingen University of Applied Sciences
Esslingen, Baden-Württemberg, Germany

ISSN 1869-8433 ISSN 1869-8441 (electronic)
Advanced Structured Materials
ISBN 978-3-031-81519-5 ISBN 978-3-031-81517-1 (eBook)
https://doi.org/10.1007/978-3-031-81517-1

This Springer imprint is published by the registered company Springer Nature Switzerland AG
The registered company address is: Gewerbestrasse 11, 6330 Cham, Switzerland

Preface

The Progress in Engineering Technology VI book features 31 research papers that collectively explore a wide range of innovative engineering solutions and advanced technologies across various fields. Most of the study focuses on designing, analyzing, and optimizing systems that improve efficiency and performance. The papers highlight cutting-edge methodologies and applications, from hydroturbine installations in underground pipelines to using nanofluids for CPU cooling. Key projects include the development of autonomous vehicles for flood detection, innovative composites for automotive applications, and advanced monitoring systems for industrial machinery. Together, these contributions tackle real-world challenges and expand the horizons of engineering practices, emphasizing the significance of interdisciplinary research in delivering impactful solutions for the future.

Kulim, Kedah, Malaysia — Tajul Adli Abdul Razak
Kulim, Kedah, Malaysia — Ahmad Kamal Ismail
Esslingen, Germany — Andreas Öchsner

Contents

Chapter 1
Design and Analysis for Hydroturbine Installation in Underground Water Pipelines

Noor Iman Fitri bin Abu Bakar, Muhammad Afif Asnawi bin Muhammad Azizan, Muhammad Rahimy bin Mohd Radzi, Wan Nur Suhada Wan Mohd Zor, and Ahamad Zaki Mohamed Noor

Abstract The water pipeline's fluid flow was utilized by the water turbine to spin the turbine runner, which ultimately produced electricity. Recently, researchers have focused more on developing a suitable water turbine design to utilize clean and renewable energy. This is because of the water delivery system's inadequate capacity which typically results in the waste of excess potential energy. Rising pollution and other environmental threats have led to a dramatic increase in the usage of renewable energy sources in recent years. The efficiency of a pipeline water turbine with a 150-mm diameter was studied using simulation, and the findings of that study are presented in this work. The examination was done to evaluate each blade design's effectiveness. To determine the best size, number, and angle of blades to be inserted in a water pipeline, the SolidWorks and ANSYS software were used to build and test the 2–6-blade turbines with 140-mm and 75-mm size turbines that are employing 2 types of angles, which are 90° and 45°. According to the data, a six-bladed turbine with blades that are 75 mm in diameter and at a 45° angle is the most suitable design

---

N. I. F. bin Abu Bakar · M. A. A. bin Muhammad Azizan · M. R. bin Mohd Radzi · W. N. S. W. M. Zor
Mechanical Engineering Section, Universiti Kuala Lumpur, Malaysian Spanish Institute Kulim Hi-Tech Park, 09000 Kulim, Kedah, Malaysia
e-mail: iman.abu23@s.unikl.edu.my

M. A. A. bin Muhammad Azizan
e-mail: afif.azizan25@s.unikl.edu.my

M. R. bin Mohd Radzi
e-mail: rahimy.radzi@s.unikl.edu.my

W. N. S. W. M. Zor
e-mail: suhada.zor@s.unikl.edu.my

A. Z. M. Noor (✉)
Manufacturing Section, Universiti Kuala Lumpur, Malaysian Spanish Institute Kulim Hi-Tech Park, 09000 Kulim, Kedah, Malaysia
e-mail: ahamadzaki@unikl.edu.my

T. A. A. Razak et al. (eds.), *Progress in Engineering Technology VI*, Advanced Structured Materials 229, https://doi.org/10.1007/978-3-031-81517-1_1

since it produces the best outlet flowrate value from impact at the incoming water flow. The findings also revealed that, with the right blade angle and number, the system's efficiency improved with increasing water flow rate. Because the size and shape of the blades generated are inappropriate, there will be turbulence formed. The efficiency, speed, and water flow of a turbine are all greatly affected by the number and design of its blades.

Keywords Hydroturbine · Blades · Renewable energy · Turbine · CFD · ANSYS

1.1 Introduction

Because it influences a turbine's speed and efficiency, the number of blades is crucial. The swept area of a wind turbine's blades directly relates to the power it generates from the wind; as a result, the blades directly impact the power generating [1]. The longer the blade, the more energy the turbine will create, but it must be balanced by at least two other objects. Costs for utilities like electricity and water could go up in the future as well. It is common knowledge that water may be used as a clean, pollution-free source of electricity. The requirement for the construction of a large dam, which has a substantial impact on the surrounding ecosystem, is the current constraint on the generation of hydrological power [2]. The management of the production process has altered because of technological advancements. The ability to outfit the clean energy industry with Internet of Things (IoT) technology has made it possible for them to link and communicate with one another to accelerate the process [3]. Recently, the world has been evolving since the 4.0 industrial revolution. Including this technology for clean energy development will be a crucial part to support any knowledge and improve progress for building up new designs and study [4].

1.2 Literature

There was a trade-off when designing turbines with more than two blades. A one-blade design is ideal for obtaining maximum efficiency due to its decreased drag. However, one blade creates an unbalance and is therefore not practicable [2]. The angle of the blades has a considerable impact on the amount of lift generated.

1.2.1 CFD Analysis

Most prior studies on crossflow water turbines measured the turbine performance using casing. The performance of the turbine can be studied prior to manufacture using computational fluid dynamics (CFD) techniques [5]. The expense and time of the experiment can be reduced by using CFD simulation. The analysis on turbine flow will combine with several design to enhance several results, by that the comparison to find the best design will be strong evidence to define the study [6]. Converting all the computation data into an animation or simulation figure is the initial stage in generating the simulation. This program uses finite volume discretization to solve the RANS equations [7]. Referring to Mani et al. [8] in CFD analysis, the geometry setup makes sure that the simulation can produce ideal results. It is necessary to set up an exit pressure and an intake mass flow rate as boundary conditions before proceeding. The author also considered the smoothness factor of the pipe wall certain value, pressure, velocity, and turbulent kinetic energy of the flow around the turbine; a prismatic mesh layer is created in the blade [9].

1.3 Methodology

Figure 1.1 shows the flowchart table for this project. To ensure that the analysis is accurate, the inner pipe diameter in this project is set at 150 mm, which is in line with the real size of the underground plumbing. Because of this, it is considered that in the internal diameter of the pipeline the hydroturbine is installed and is less than 150 mm. In this project, 20 different types of turbines were designed.

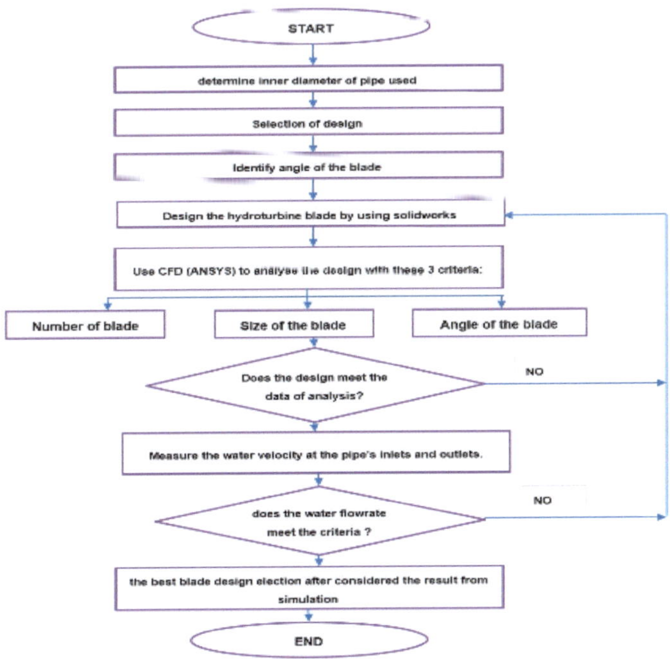

Fig. 1.1 Flowchart of work for design and analysis of hydroturbine

1.3.1 Water Flowrate at Inlet and Outlet of the Pipe

Since the velocity of the flowrate should be set according to Water Authorities' and Malaysian Water Associate's (MWA) guidelines and "Pengurusan Aset Air Berhad" (PAAB) requirement, this value should not exceed 2.0 m/s to 2.5 m/s. The water's speed in this project will be lower than the required velocity. The inlet velocity has been set as by a constant water velocity profile with 1.5 m/s.

1.4 Result and Discussion

The simulations of water flow through turbines of diameters 140 and 75 mm generated the observed outcomes. The inlet velocity remained constant throughout the experiment at 1.5 m/s. The inlet and outflow contour profiles and graph were generated with the ANSYS Fluent software. In this study, it was hypothesized that a greater total number of water turbine blades would result in greater overall system efficiency. Below is the result representing three parameters

1.4.1 Overall Result Discussion

Moving to Table 1.1 with 75-mm diameter of blade, this blade has better water flow with less water turbulence at the turbine. The reading at the outlet is not too far for all regions where the lower reading is not too slow (0.000e-00–4.199e-01 m/s) and (2.240e + 01–2.520e + 01 m/s) at higher reading region. The differences reading at lower and higher are not too far as blade with 140 mm; hence, the blade size of 75 mm is selected as the best size to install in water pipeline for these cases. For this experiment, it was discovered that smaller blades in the runner result in improved turbine performance. This is shown in Table 1.1. From the graph in size 140 mm, it shows unsteady and inconsistent water flow after the water flows through the blade. The reading of this blade is not stable since there is very slow water flow in blue color that produced a turbulence flow (0.000e-00–8.027e-01 m/s) and some region is in red color (4.28e + 01–4.816e + 01 m/s) which with slightly higher velocity is shown in the velocity contour.

Table 1.2 shows how a turbine's total efficiency changes as the number of its blades increases. As a turbine's blade count increases, its efficiency tends to rise. From the simulation result of all blade design, the performance of turbine will increase if a smaller and proper size of blade is used. The best size in this case is 75 mm. In addition, if the turbine is tested by using a smaller number of blades compared to a greater number of blades, the water flow will be minimal due to turbulence impact because of high rotation. Above all the results and discussions, the performance of the blade number is best achieved by the 6-blade design with size of 75 mm and angle 45°.

Table 1.1 Comparison of different size of blade

Size	Parameter	Contour/graph effects of size of blades on water flow
140 mm	Velocity contour and graph for: angle: 45° No. of blade: 6	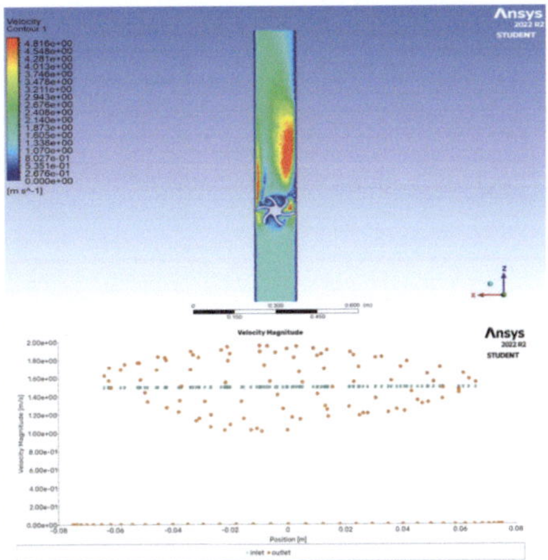
75 mm	Velocity contour and graph for: angle: 45° No. of blade: 6	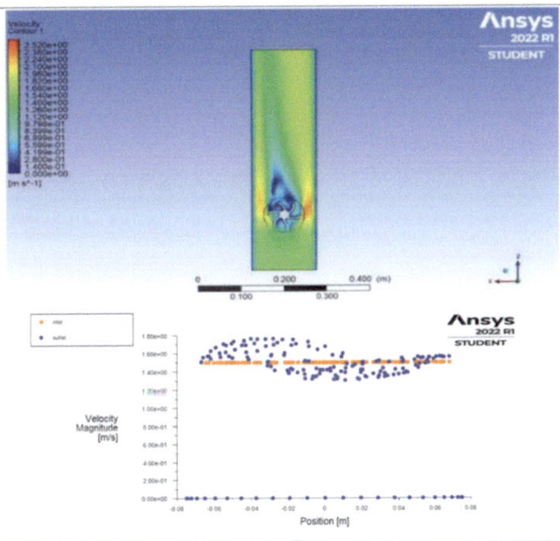

1.4.2 *Best Blade Design*

From the simulation result of all blade design, the performance of turbine will increase if a smaller and proper size of blade is used. The best size in this case is 75 mm. In addition, if the turbine is tested by using a smaller number of blades compared to a greater number of blades, the water flow will be minimal due to turbulence impact because of high rotation. Above all the results and discussions, the performance of the blade number is best achieved by the 6-blade design with size of 75 mm and angle 45°.

Table 1.2 Comparison between different number of blades

Number of blades	Parameter	Contour/graph effects of number of blades on water flow
4	Velocity contour and graph for: angle: 45° size of blade:75 mm	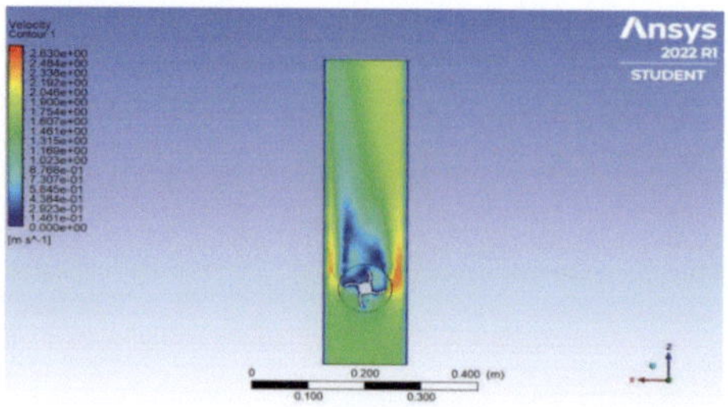 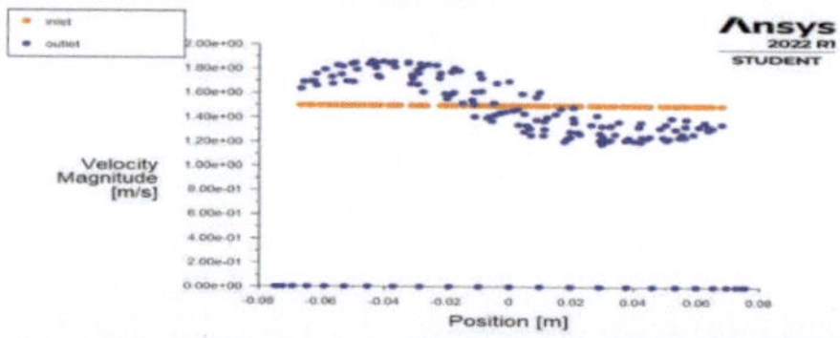
6	Velocity contour and graph for: angle: 45° size of blade:75 mm	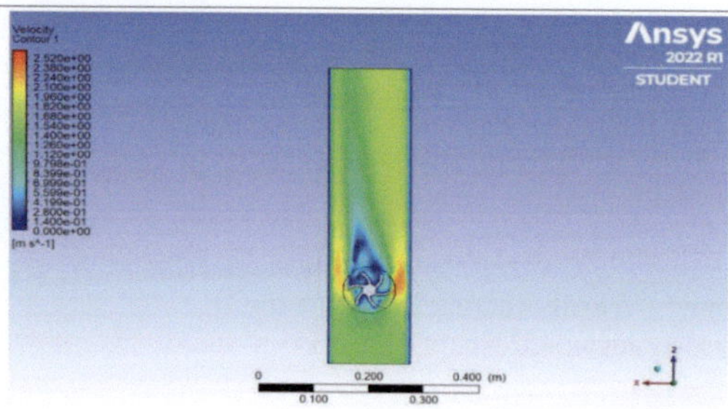 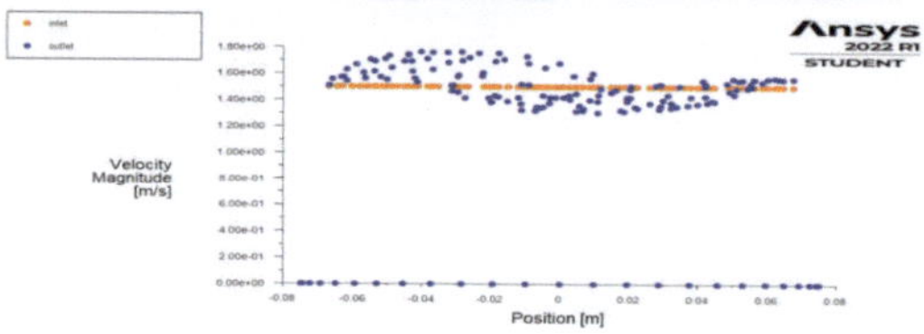

1.5 Conclusions

Turbines with less than five or six blades are less effective because they are unable to utilize the full potential of the flow capacity. When the hydroturbine is set up with 6 blades, which performs better than the 2, 3, 4, and 5 blades, the best turbine efficiency is discovered. A water turbine with this blade will be appropriate for further development. The study also discovered that when installed in a pipeline, curved blades perform better than straight blades.

References

1. Adeyeye KA, Ijumba N, Colton J (2021) The effect of the number of blades on the efficiency of a wind turbine. In: IOP conference series: earth and environmental science, vol 801(1). IOP Publishing, p 012020
2. Sritram P, Suntivarakorn R (2019) The effects of blade number and turbine baffle plates on the efficiency of free-vortex water turbines. In: IOP conference series: earth and environmental science 2019 Apr 1, vol 257(1). IOP Publishing, p 012040
3. Fauadi MHFM, Damanhuri AAM, Hambali RH, Mohamed AZ, Noor NIA (2020) Recent applications of internet of things (IoT) in manufacturing sector–a review. Int J Sci Technol Res 9(09):334–339
4. Noor AZM, Kamil MZM, Hisban NHNM, Anuar NNM, Khairudin SNM, Jafar FA (2023) Development of an IoT-smart parking mall sensor using blynk and ThingSpeak. In: Abu Bakar MH, Abdul Razak TA, Öchsner A (eds) IT applications for sustainable living. SpringerBriefs in applied sciences and technology. Springer, Cham, pp 225–34 (2023)
5. Acharya N, Kim C-G, Thapa B, Lee Y-H (2015) Numerical analysis and performance enhancement of a crossflow hydro turbine. Renew Energy 819–826
6. Semerci DS, Yavuz T (2017) Increasing efficiency of an existing Francis turbine by rehabilitation process. In: 2016 International conference on renewable energy research and applications (ICRERA). IEEE, pp 107–111
7. Günel O, Koç E, Yavuz T (2019) CFD vs. XFOIL of airfoil analysis at low Reynolds numbers. In: 2016 IEEE international conference on renewable energy research and applications (2019)
8. Mani et al (2019) Effect of changing number of blades and discharge on the performance of a cross-flow turbine for micro hydro power plants. Int J Eng Sci Comput 6(8)
9. Aziz MS, Khan MA, Jamil H, Jamil F, Chursin A, Kim DH (2021) Design and analysis of in-pipe hydro-turbine for an optimized nearly zero energy building. Sensors 21(23):8154

Chapter 2
Optimal Tensile Properties of Carbon Composite Made of Empty Fruit Brunch Using Taguchi Method

Mior Asyraf Al Hafiz Bin Mior Izzuddin, Mohamad Sazali Bin Said, Mohd Shahrizan Bin Yusoff, and Mohd Zaki Bin Abdul Razak

Abstract Industrial pollution, especially in palm oil, has increased in recent years. This has raised awareness of the need for biodegradable material research, such as the carbon-reinforced empty fruit bunch (EFB) carbon composite. Researchers used the Taguchi method to assess the mechanical properties of the EFB carbon composite. Tensile strength, modulus of elasticity, and elongation were investigated and how weight % and speed affect them. A 3D printer was utilized to make the mold for this research using ASTM D638-14 standards. Taguchi-recommended EFB carbon composite compositions of 30, 40, and 50% were used to make specimens. Tensile testing was done with a Universal Testing Machine (UTM) and evaluated using Taguchi method L9. The investigation found that 40% weight and 6 mm/min speed were appropriate for EFB carbon composite. The composite has 3.7 MPa tensile strength, 223.149 N/m^2 modulus of elasticity, and 26.9% elongation. This was compared to 30 and 50% weight percentage, 5 mm/min and 7 mm/min outcomes. The findings of this study can aid green material research and development.

Keywords Empty fruit bunch · Mechanical properties · Taguchi method

---

M. A. Al Hafiz Bin Mior Izzuddin · M. S. B. Said (✉) · M. S. B. Yusoff · M. Z. B. A. Razak
Universiti Kuala Lumpur, Malaysian Spanish Institute Kulim Hi-Tech Park, 09000 Kulim, Kedah, Malaysia
e-mail: msazali@unikl.edu.my

M. A. Al Hafiz Bin Mior Izzuddin
e-mail: mior.mior10@s.unikl.edu.my

M. S. B. Yusoff
e-mail: mshahrizan@unikl.edu.my

M. Z. B. A. Razak
e-mail: mzaki@unikl.edu.my

T. A. A. Razak et al. (eds.), *Progress in Engineering Technology VI*, Advanced Structured Materials 229, https://doi.org/10.1007/978-3-031-81517-1_2

2.1 Introduction

Pollution is the introduction of substances or factors into the environment that are dirty, unsanitary, or harmful, resulting in a detrimental impact on the environment. In addition, pollution can pose risks to both human health and the environment. There exist several forms of pollution, encompassing natural occurrences, such as earthquakes and pollution, resulting from human actions [1].

Over the past several years, there has been a rise in environmental awareness and concern over global warming. As a result, scientists worldwide have been adopting and utilizing environmentally friendly goods. The rising popularity of eco-friendly materials may be attributed to their affordability, ecologically conscious properties, ease of implementation, and impressive effectiveness [2]. Due to their lower cost, greater cost-effectiveness, and superior environmental impact, natural fibers are the preferred choice over synthetic and artificial fibers.

Researchers are increasingly focusing on biodegradable plant fibers, such as empty fruit bunch, which are now utilized in the automotive, industrial, and construction sectors. By transforming EFB into a product like activated carbon (AC), we may effectively address environmental concerns by repurposing the trash and generating a valuable resource for another sector.

This study aims to investigate the utilization of empty fruit bunch (EFB) carbon as a primary component in the production of composites, in conjunction with epoxy resin and hardener. The aim is to evaluate the mechanical characteristics of the composite material prior to its utilization in manufacturing. A Universal Testing Machine (UTM) will be utilized to conduct a tensile test in order to ascertain the tensile strength, elastic modulus, and elongation. The Taguchi method will be employed to examine the data and ascertain the ideal outcomes of the composite. The study aims to examine the potential impact of weight % and speed on the mechanical characteristics of EFB carbon composites.

2.2 Methodology

The tensile test mold was designed and fabricated in SolidWorks to achieve the objective. The 3D printing mold is made according to ASTM 368–14, as shown by the tensile specimen schematic in Fig. 2.1. According to ASTM D638-14 standards, a dumbbell-shaped test specimen with an overall length of 165 mm and a thickness of 3.2 0.4 mm will be produced. Although the entire width is 19 mm, the overall radius is 76 mm [3].

The material to be utilized is a combination of EFB carbon composite, epoxy resin, and hardener. As per the Taguchi method outlined in Minitab software, a total of nine sample specimens have been created for the purpose of testing. An EFB carbon composite was created by adjusting the amount of activated carbon to 30%, 40%, and 50%.

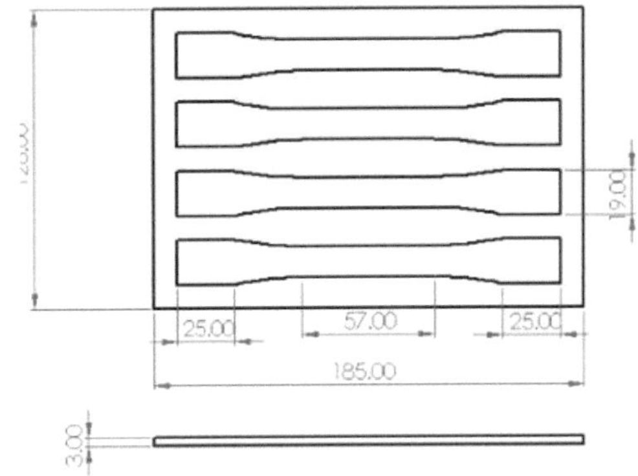

Fig. 2.1 Schematic drawing on SolidWork

Specimens 1, 2, and 3 require a mixture consisting of 30% EFB carbon and 70% epoxy resin. However, specimens 4, 5, and 6 necessitate a combination of 40% EFB carbon and 60% epoxy resin. Specimens 9, 8, and 9 necessitate a combination of 50% EFB carbon and 50% epoxy resin [4]. The carbon content of the empty fruit bunch (EFB) was determined using a micro-weighing scale, following the values provided in Table 2.1 for the composition of the EFB carbon composite and Table 2.2 for the composition of the empty fruit bunch composite, both measured in grams.

The composite mixtures will be left at room temperature, typically the temperature between 20 and 25 degrees Celsius, for 24 h until they have hardened. Figure 2.2 shows the EFB carbon composite specimen 1 to specimen 9 according to ASTM 638. After 24 h, carefully remove the EFB carbon composites from the 3D printing mold. After that, the specimens will be test finished with a tensile test using Universal Testing Machine (UTM).

After the EFB carbon composite specimens were fabricated, they were subjected to mechanical tests according to ASTM standards. These tests evaluated the mechanical properties of the EFB carbon composite specimens. The test evaluates the tensile

Table 2.1 Empty fruit bunch composites composition L9 orthogonal array

Number of specimens	Empty fruit bunch carbon, (%)	Epoxy resin and hardener, (%)	Total, (%)
1	30	70	100
2	30	70	100
3	30	70	100
4	40	60	100
5	40	60	100
6	40	60	100
7	50	50	100
8	50	50	100
9	50	50	100

Table 2.2 Empty fruit bunch composite composition in gram

Number of specimen	EFB carbon, (g)	Epoxy resin, (g)	Hardener, (g)	Total, (g)
1	3	5.25	1.75	10
2	3	5.25	1.75	10
3	3	5.25	1.75	10
4	4	4.50	1.50	10
5	4	4.50	1.50	10
6	4	4.50	1.50	10
7	5	3.75	1.25	10
8	5	3.75	1.25	10
9	5	3.75	1.25	10

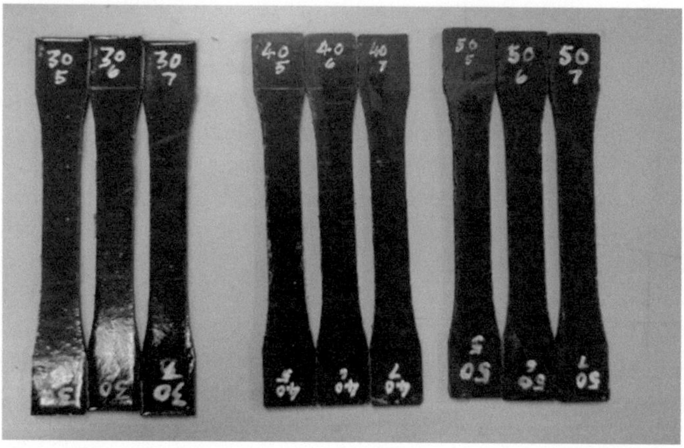

Fig. 2.2 EFB carbon composite specimen 1 to specimen 9 according to ASTM 638

strength, modulus of elasticity, and elongation. The tensile test was conducted at different speeds to evaluate the mechanical properties of the EFB carbon composite specimen.

2.3 Result and Discussion

To analyze the mechanical properties of the tensile strength, modulus of elasticity, and elongation, tensile test has conducted with different weight percentage and speed. Table 2.3 shows results of tensile test using Universal Tensile Machine (UTM).

Table 2.3 Results of tensile test using universal tensile machine (UTM)

Experiment number	Parameter		Tensile strength (MPa)	Modulus elasticity (N/mm^2)	Elongation %
	Weight percentage %	Speed (mm/min)			
1	30	5	2.655	83.14	31.38
2	30	6	3.878	316.769	23.68
3	30	7	3.578	251.015	39.54
4	40	5	4.496	388.817	15.9
5	40	6	3.7	223.149	26.9
6	40	7	2.507	194.811	12.87
7	50	5	2.938	238.657	8.58
8	50	6	2.514	148.063	8.92
9	50	7	2.464	126.206	10.54

2.3.1 Tensile Strength

The material to be utilized consists of a combination of EFB (empty fruit bunch) carbon composite, epoxy resin, and hardener. According to the Taguchi method, as outlined in Minitab software, a total of nine sample specimens have been created for testing purposes. The carbon composite material, made up of EFB and activated carbon, was prepared by adjusting the amount of activated carbon to 30%, 40%, and 50%.

Specimens 1, 2, and 3 were prepared with a mixture consisting of 30% EFB carbon and 70% epoxy resin. Specimens 4, 5, and 6 required a composition of 40% EFB carbon and 60% epoxy resin. Lastly, specimens 7, 8, and 9 necessitated a combination of 50% EFB carbon and 50% epoxy resin. Initially, the carbon content in the EFB was quantified using a micro-weighing scale, referring to the data provided in Table 2.1 for the composition of the EFB carbon composite and Table 2.2 for the composition of the empty fruit bunch composite, both measured in grams (Table 2.4).

Figure 2.3 shows the graph of tensile strength that was analyzed by Minitab software. The optimum point explained producing the higher tensile strength. For tensile strength, weight percentage is 40%, while the speed is 6 mm/min for obtaining the higher tensile strength.

Table 2.4 ANOVA for S/N ratios of tensile strength

Source	DF	Seq SS	Adj SS	Adj MS	F	P
Weight percentage %	2	10.061	10.061	5.031	1.19	0.394
Speed (mm/min)	2	3.805	3.805	1.902	0.45	0.667
Residual error	4	16.951	16.951	4.238		
Total	8	30.818				

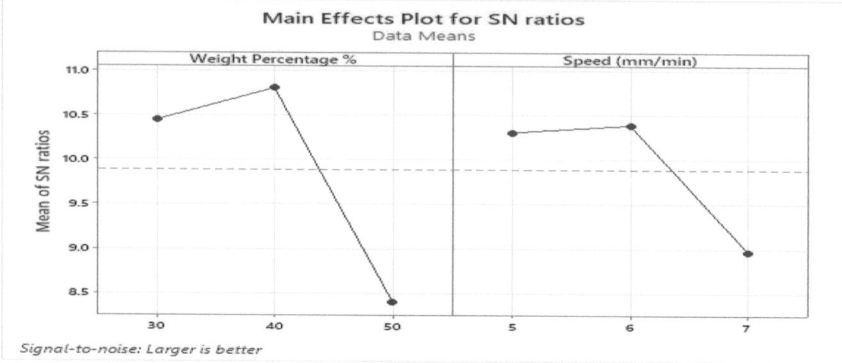

Fig. 2.3 Tensile strength graph

Agreed with [5], it has been reported that oil palm fiber is not robust enough to withstand heavy loads, which leads to the failure of the fiber and subsequently results in a lower flexural strength of pure oil palm empty fruit bunch (EFB) composites. This means that the pure oil palm EFB composites may not be suitable for applications that require a high level of strength and durability. This highlights the need for further research and development to improve the mechanical properties of oil palm fiber and EFB composites.

Besides that [6] reported, with increasing oil palm EFB in hybrid composites, the flexural properties decreased. Oil palm EFB fibers display poor stress transmitted from the polymer matrix due to its structure and shape. To get the strengths of EFB carbon composite specimens, the bonding between EFB carbon composite and matrix must be strong.

Furthermore, the weight percentage and speed in EFB carbon composite does not have a significant effect on the tensile strength. These related with null hypothesis earlier that there is no significant relationship between weight percentage and speed.

2.3.2 *Modulus Elasticity*

Modulus elasticity is a quantitative measure of the stiffness of a material. As the stiffness of materials increases, the amount of deformation that occurs under loading decreases. The analysis of variance (ANOVA) conducted on the tensile test results revealed that the modulus of elasticity is not significantly affected by the weight % (p-value $= 0.679$) and speed (p-value $= 0.939$). These parameters are considered insignificant since their values are more than 0.05. Modifying the weight % or speed will have no impact on the modulus of elasticity. Table 6 displays the analysis of variance (ANOVA) results for the signal-to-noise (S/N) ratios of modulus elasticity.

After doing a thorough analysis of the data, it was determined that the null hypothesis, which states that there is no significant correlation between weight % and speed, cannot be disproven. Thus, it can be concluded that the weight % of an object does not exert a substantial impact on its speed in this investigation (Table 2.5).

Figure 2.4 shows the graph of modulus elasticity that was analyzed by Minitab software. The optimum point explained producing the higher modulus elasticity. For modulus elasticity, weight percentage is 40%, while the speed is 6 mm/min for obtaining the higher modulus elasticity.

The maximum modulus of elasticity, as depicted in Fig. 2.4, was determined to be 388.817 N/m^2. This measurement was achieved with a weight percentage of 40% and a testing speed of 5 mm/min. It has been observed that the modulus of elasticity of the EFB carbon composite tends to decrease with an increase in weight percentage and a decrease in testing speed. The modulus of elasticity is influenced by both the weight percentage and the strength of the adhesive bonding at the interface. Poor interfacial bonding is identified as a primary cause of the reduction in the modulus of elasticity. Additionally, the quantity of the matrix material also plays a significant role in enhancing the modulus of elasticity, thereby improving the composite's ability to withstand external forces.

Previous research has similarly indicated that increasing the amount of oil palm empty fruit bunch (EFB) in an epoxy polymer matrix adversely affects the tensile strength and tensile modulus of the resultant composite [7]. These findings align with the null hypothesis, which posits no statistically significant correlation between

Table 2.5 ANOVA for S/N ratios of modulus elasticity

Source	DF	Seq SS	Adj SS	Adj MS	F	P
Weight percentage %	2	23.568	23.568	11.784	0.43	0.679
Speed (mm/min)	2	3.536	3.536	1.768	0.06	0.939
Residual error	4	110.338	110.338	27.584		
Total	8	137.442				

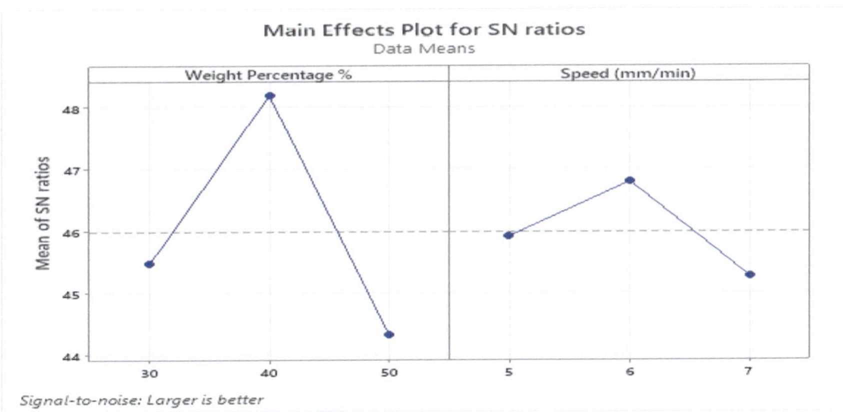

Fig. 2.4 Modulus elasticity graph

weight percentage and testing speed. Therefore, increasing the amount of oil palm EFB in the epoxy polymer may not be the most effective method for improving the mechanical properties of the composite. Further research is necessary to identify alternative methods for enhancing the strength and durability of oil palm EFB composites.

2.4 Conclusions

The findings indicate that the ideal parameters were achieved with a weight percentage composition of 40% and a speed of 6 mm/min. These parameters resulted in a tensile strength of 3.7 MPa, an elasticity modulus of 223.149 N/m^2, and an elongation of 26.9%. The goal is to determine the ideal parameters of EFB carbon composite by employing the Taguchi technique, which will be accomplished after the desired mechanical characteristics of the EFB carbon composite specimens are attained. The weight % and speed do not significantly affect the tensile strength and modulus of elasticity, as indicated by the p-value being more than 0.05. The result for elongation is statistically significant, since the p-value is less than 0.05. An analysis has been conducted on the mechanical characteristics of EFB carbon composite. Based on the Taguchi study, we have concluded that there is no statistically significant impact of weight % and speed on the tensile strength and modulus of elasticity of EFB carbon composite. Based on the Taguchi analysis, we have concluded that there is a considerable influence on the weight % and speed for elongation, therefore accepting the alternative hypothesis. In summary, the research suggests that more enhancements are necessary before the EFB carbon composite may be effectively utilized in the industrial sector.

References

1. Onoja E, Chandren S, Abdul Razak FI, Mahat NA, Wahab RA (2019) Oil palm (Elaeis guineensis) biomass in Malaysia: the present and future prospects. Waste Biomass Valorization 10(8):2099–2117
2. Ozgul EO, Ozkul MH (2018) Effects of epoxy, hardener, and diluent types on the workability of epoxy mixtures. Constr Build Mater 158:369–377
3. Ramanaiah K, Prasad AVR, Chandra KH (2012) Effect of fiber loading on mechanical properties of borassus seed shoot fiber reinforced polyester composites. J Mater Environ Sci 3(3):374–378
4. Samotu IA, Dauda M, Anafi FO, Obada DO (2015) Suitability of recycled polyethylene/palm kernel shell-iron filings composite for automobile appliction. Tribol Ind 37(2):142–153
5. Sastry PK (2017) An overview of ceramic matrix composites. Int J Adv Eng Res Dev 4(05):23–26
6. Shandilya A, Gupta A, Verma D (2016) Banana fiber reinforcement and application in composites: a review. Green Approaches Biocomposite Mater Sci Eng 201–227
7. Tanaka R, Yamamoto K (2004) Accumulation and availability of oil palm biomass in Malaysia. JIRCAS Working Report, vol 39, pp 20–34. https://www.cabdirect.org/cabdirect/abstract/20083249030

Chapter 3
Investigation of Al_2O_3 Nanofluid Performance for Computer Central Processing Unit Cooling Systems

Nur Anisah Mohd Jamil, Ahmad Kamal Ismail, Khairul Akmal Shamsuddin, Mohd Fauzi Abu Hassan, Nor Haslina Ibrahim, Mohamad Shukri Mohd Zain, and Ayub Ahmed Janvekar

Abstract The central processing unit (CPU) is the most critical component of the computer's hardware. During operation, the CPU produces heat, just like any other electronics devices, and it must be effectively cooled in order to operate at the peak best performance. In this work, alumina (Al_2O_3) nanofluid mixed water has been used as a cooling system for the CPU. Nanofluid refers to a working fluid that has nanoparticles mixed in a base fluid, such as water or ethylene glycol. To improve the current cooling system, the coolant fluid has been changed from distilled water to a nanofluid coolant. The main focus of this work is to design and modify the water cooling system and to test the performance of the nanofluid cooling system for the computer CPU. It has been found from the experiment testing that the water cooling

N. A. M. Jamil · A. K. Ismail (✉) · K. A. Shamsuddin · M. F. A. Hassan · N. H. Ibrahim · M. S. M. Zain
Universiti Kuala Lumpur Malaysian Spanish Institute, Kulim Hi-Tech Park, 09000 Kulim, Kedah, Malaysia
e-mail: ahmadkamal@unikl.edu.my

N. A. M. Jamil
e-mail: nanisah.jamil@s.unikl.edu.my

K. A. Shamsuddin
e-mail: khairulakmal@unikl.edu.my

M. F. A. Hassan
e-mail: mohdfauzi@unikl.edu.my

N. H. Ibrahim
e-mail: norhaslina@unikl.edu.my

M. S. M. Zain
e-mail: mshukrimz@unikl.edu.my

A. A. Janvekar
School of Mechanical Engineering, Vellore Institute of Technology, Chennai, Tamil Nadu 600127, India
e-mail: ayubahmed.janvekar@vit.ac.in

T. A. A. Razak et al. (eds.), *Progress in Engineering Technology VI*, Advanced Structured Materials 229, https://doi.org/10.1007/978-3-031-81517-1_3

system improved by using a nanofluid by 2.39%. The overall heat transfer coefficient between those two coolants has been calculated and compared. The overall heat transfer coefficient of nanofluid mixture is higher than for pure distilled water. The smart control system also provides a good monitoring system to the CPU working temperature. In conclusion, a nanofluid can be the best option for a CPU cooling system.

Keywords CPU · Water cooling system · Alert system · Nanofluid · Aluminum oxide · Overall heat transfer coefficient · Heat transfer performance

3.1 Introduction

In general, the computer central processing unit (CPU) is generating heat during operation, and therefore, it must be effectively cooled to optimize its performance. Nowadays, nanofluid coolants have been widely used for cooling systems by mixing them with water. A nanofluid refers to a working fluid that has nanoparticles mixed in a base fluid, such as water or ethylene glycol. To improve the CPU cooling system, the air cooling system can be replaced with a water cooling system, while the distilled water with an air-cool blower system can be substituted with an Al_2O_3-based nanofluid mixture. Furthermore, a proper design of the water cooling system needs to be done to suit the compact space and limited area in the CPU.

Previous work [1] has been found which demonstrates a novel cooling approach utilizing a combination of heat sinks and a water system as a more efficient alternative to the conventional air cooling system. The issue is that ineffective cooling can reduce the server's reliability because of hardware exceeding its stated temperature limits. The goal of this study is to redesign the heat sink using heat pipes to better transmit heat to the water channels. The method adopted was to build a new system that is more environmentally friendly and economically effective. The next step is to design a smart Android-based temperature control model that can be used to change the cooling system's water flow rate and to remotely monitor the CPU's thermal status. As a result, a cooling method employing both heat sinks and water was developed, as well as an Android-based smart thermal control and remote monitoring control system. Hence, the proposed cooling system's total efficiency is greatly increased when compared to other traditional cooling systems.

There is a fact that microfluidic technology has been enhanced to include monitoring, temperature control, and cooling sections [2]. The issue with this project is that the CPU becomes overheated, thus slowing down the machine's operations. Nowadays, computers depend entirely on a cooling fan or heat sink to cool the processor. The purpose is to create a microfluidic fluid flow system with a temperature control system. The next step is to monitor the processor's temperature using an intelligent temperature controller (XHW1401), control the system's temperature using an intelligent temperature controller (XHW1401), and take over the entire system using an Arduino relay. The method used in this project was to develop the controller using an Arduino and the Blynk app. As the result, when the processor becomes too hot, the temperature of the system with the microfluidic microfin will send a notification about the condition of CPU and the temperature of the block will be decreased.

According to Nguyen et al. [3], an experiment was conducted to determine the behavior and enhancement of heat transfer of a specific nanofluid, an Al_2O_3 nanoparticle–water mixture, moving inside a closed system intended for cooling microprocessors or other electronic components. The issue is that there is a shortage of data on the behavior of nanofluids in actual thermal applications. To address this issue, the authors perform a heat transfer experiment using a typical liquid cooling system using a nanofluid composed of distilled water and Al_2O_3 nanoparticles with two different particle average diameters and various volume concentrations in place of the base fluid. Experiments have demonstrated clearly that the nanofluid with a smaller particle diameter has higher heat transfer coefficients than the nanofluid with a larger particle size. The overall heat transfer coefficient (OHTC) measurement method can be found in a previous work [4–7]. Nanofluids and hybrids are well-known working fluids with a high heat transfer coefficient in heat transfer applications.

3.2 Methodology

The component of the water cooling system has been designed using SolidWorks (see Fig. 3.1), and the details of the nanofluid used are shown in Fig. 3.2. The capacity of the mixture of distilled water and nanofluid, which is aluminum oxide, is 17.3 L for the distilled water and 0.2 L for the nanofluid. Nanofluid accounts for 1.156% of the total volume of the mixture. The overall heat transfer coefficient is increased using the nanofluid mixture.

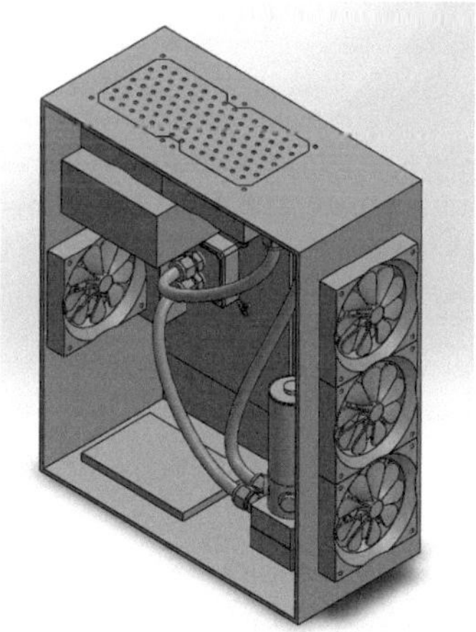

Fig. 3.1 Assembly design of water cooling system

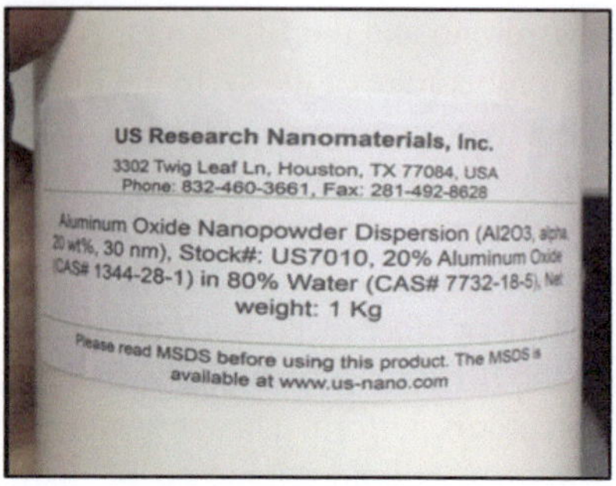

Fig. 3.2 Mixture of distilled water and aluminum oxide

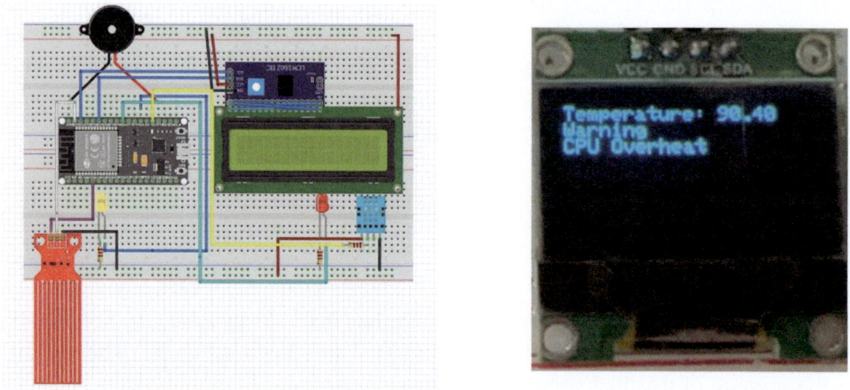

Fig. 3.3 Circuit wiring (left) and OLED indicator (right)

3.2.1 Circuit Diagram and OLED Display for the Alert System

A water level sensor was used to detect water, a DHT 11 to collect the CPU's temperature, an OLED to display the system's status, a buzzer, and a LED to function as output, and the ESP32 as the main microcontroller. There are many sensors available in the market. The one used in this project is shown in Fig. 3.3. Other information regarding the control and sensor can be found from the work by Ibrahim et al. [5].

3.2.2 Overall Heat Transfer Coefficient

The total heat transfer coefficient, often known as the *U*-value, measures the heat transmission through a material or system and is typically expressed in units of $W/m^2{\cdot}K$ (watts per square meter per degree Kelvin). A lower *U*-value indicates that the material or system transfers less heat, while a larger *U*-value indicates that more heat is transferred. The overall heat transfer coefficient can be obtained by finding the heat transfer, $Q = \dot{m}\, c_p \Delta T$ followed by the overall heat transfer coefficient, $U = Q\,/\,A\,\Delta T$.

3.3 Result and Discussion

3.3.1 *Experimental Results*

Before changing the water cooling system for the CPU, the previous cooling system that has been used was an air cooling system. After changing the system, two experiments were conducted for the water cooling system using distilled water without nanofluid, and for the second experiment, it is a nano mixture with distilled water.

The temperature for the water cooling system using the nanofluid mixture has the lowest temperature compared to the air cooling system and water cooling system using distilled water (see Fig. 3.4). The average of overall heat transfer coefficient for distilled water is 110.442 kW/m^2.K while for the nanofluid mixture with distilled water is 113.091 kW/m^2.K. The overall heat transfer coefficient for the nanofluid mixture is higher than for distilled water. The higher the overall heat transfer coefficient, the better the heat transfer performance.

The increment percentage for overall heat transfer coefficient (OHTC) for a nanofluid mixture of aluminum oxide with distilled water will vary depending on the specific conditions and parameters of the system. Studies have shown that the use of nanofluids can increase the OHTC by as much as 40–60% (see Fig. 3.5) compared to traditional fluids, but it is important to note that these results can be affected by factors, such as nanoparticle concentration, temperature, and flow rate. This experiment was conducted to get a specific percentage increment for the nano mixture

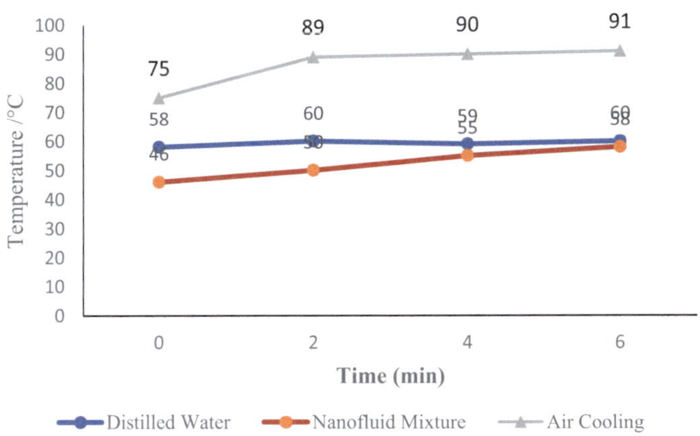

Fig. 3.4 Temperature versus time for different coolants

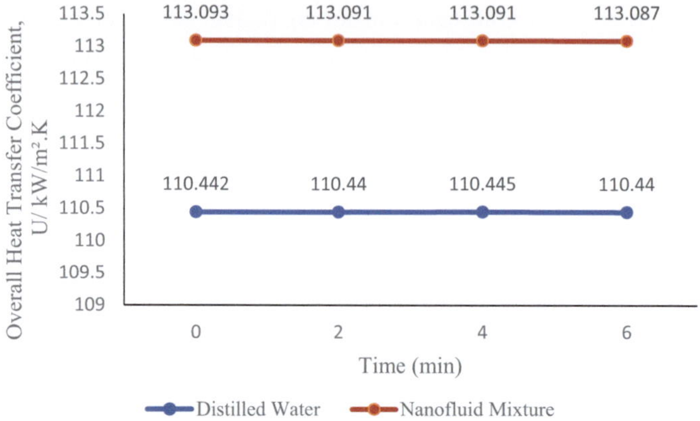

Fig. 3.5 Overall heat transfer coefficient of distilled water versus nanofluid mixture

aluminum oxide with distilled water. In this study, after using 1.156% of aluminum oxide, there is 2.39% of increment for the overall heat transfer coefficient compared to distilled water.

3.4 Conclusions

In conclusion, this project highlights to design and development of a nanofluid cooling system for a computer CPU. Data from the experimental study has been collected and discussed in previous sections. Designing the open loop water cooling system using SolidWorks plays a significant role before fabricating the actual open loop water cooling system because it can show the suitable design to fabricate. After designing the system, an alert system has been developed by using a microcontroller and integrated with the water cooling system. The microcontroller is ESP32. A water level sensor and a temperature sensor are used to send signals to the microcontroller. The distilled water was replaced with a nanofluid mixture which is aluminum oxide. The experiment between distilled water and the nanofluid mixture with distilled water has been done. The overall heat transfer coefficient between those two coolants has been calculated and compared. The overall heat transfer coefficient of the nanofluid mixture is higher than distilled water. Based on the result, the nanofluid mixture shows an increment of 2.39% in terms of overall heat transfer coefficient.

References

1. Amin M, Abdel Aziz GA, Naraghi M, Palatty M, Benz A, Ruiz D (2020) Improving the data center servers cooling efficiency via liquid cooling-based heat pipes. In: 2020 , IAS 2020, pp 3–7
2. Connor N (2019) What is overall heat transfer coefficient—U-factor—definition, thermal engineering. Available at: https://www.thermal-engineering.org/what-is-overall-heat-transfer-coefficient-u-factor-definition/. Accessed 14 Jan 2023
3. Martínez-Merino P, Sani E, Mercatelli L, Alcántara R, Navas J (2020) ACS Sustain Chem Eng 8(3):1627–1636. https://doi.org/10.1021/acssuschemeng.9b06489
4. Ch5—cooling of power semiconductor devices PDF (no date) Scribd. Scribd. Available at: https://www.scribd.com/document/429178411/Ch5-Cooling-of-Power-Semiconductor-Devices-pdf. Accessed 14 Jan 2023
5. Manan ANA, Anuar D, Khalid MLA, Ismail AK, Azid IA (2022) Overall heat transfer coefficient of different coolants and frontal air velocity in automotive radiators. In: Progress in engineering technology IV. Springer, pp 139–146, 2022
6. Harun MA, Sidik, NAC (1970) A review on development of liquid cooling system for central processing unit (CPU), Mendeley. Available at: https://www.mendeley.com/catalogue/eaafde68-753d-35b2-8302-0b1cd75e7dcd/?utm_source=desktop&utm_medium=1.19.8&utm_campaign=open_catalog&userDocumentId=%7B9c16c3d3-fa0b-428b-8bfd-834e76d53ebb%7D. Accessed 14 Jan 2023
7. Ibrahim NH, Baharuddin B, Ismail AK, Shamsuddin KA (2023) A smart control of solar panel mechanism with arduino based data logger for neural network analysis. Politek Kolej Komuniti J Eng Technol 8(1):1–12

Chapter 4
Noise Absorption and Tensile Properties of Hybrid Waste Rubber Tyre-Based Polypropylene Composites for Car Door Panel Applications

Siti Rohana Binti Ahmad, Muhammad Akram Bin Zunkipeli, Mohamad Sazali Bin Said, and Muhammad Zunnurrin Bin Ghazali

Abstract This research explores the use of hybrid waste rubber tyre-based polypropylene and coconut fibre composites for car door panel applications, aiming to balance noise absorption and mechanical strength. The composite was prepared using mixing and hot press-moulded machines, with coconut fibre content increasing tensile strength and elongation at break. The composite had the highest tensile strength of 5.25 MPa, but the tensile strength decreased with increasing coconut fibre content. Young's modulus increased by 62% but reduced elongation at break by 14.71%. The impendence tube test showed a higher sound transmission loss for this new composite compared to the current material.

Keywords Noise · Sound absorber · Waste tyre · Polypropylene · Fibre · Impendence tube

S. R. B. Ahmad (✉) · M. A. B. Zunkipeli
Mechanical Section, Universiti Kuala Lumpur, Malaysian Spanish Institute Kulim Hi-Tech Park, 09090 Kulim, Kedah, Malaysia
e-mail: sitirohana@unikl.edu.my

M. A. B. Zunkipeli
e-mail: akram.zunkipeli@s.unikl.edu.my

S. R. B. Ahmad
Centre for Women Advancement and Leadership, Universiti Kuala Lumpur, 50250 Kuala Lumpur, Wilayah Persekutuan, Malaysia

M. S. B. Said
Manufacturing Section, Universiti Kuala Lumpur, Malaysian Spanish Institute Kulim Hi-Tech Park, 09090 Kulim, Kedah, Malaysia
e-mail: msazali@unikl.edu.my

M. Z. B. Ghazali
R&Z Auto, 470 Jalan Kota Kenari 3, 09000 Kulim, Kedah, Malaysia

T. A. A. Razak et al. (eds.), *Progress in Engineering Technology VI*, Advanced Structured Materials 229, https://doi.org/10.1007/978-3-031-81517-1_4

4.1 Introduction

The automotive industry is striving for sustainable solutions to improve vehicle performance and environmental sustainability. Car door panels, crucial for aesthetics and functionality, often struggle with balancing noise insulation and mechanical strength [1, 2]. Road noise is a significant concern, affecting interior comfort and acoustic quality. To ensure a quieter cabin, panels must effectively dampen and block noise, preventing a less enjoyable driving experience and increased passenger fatigue [1, 3, 4].

Polypropylene (PP) is a versatile thermoplastic polymer with numerous industrial applications due to its chemical resistance, low density, and durability. Its flexibility, heat resistance, and moldability make it ideal for packaging, automotive components [5, 6], consumer goods, and medical devices. Despite its cost-effectiveness and environmental benefits, PP continues to be a fundamental material in various industries [7–9].

Waste tyre-based materials, such as rubber composites, can be used in car door panels to enhance their noise insulation properties [10]. The rubber's porous structure helps absorb and dampen sound waves, reducing the transmission of road noise into the cabin. Additionally, utilizing waste tyre materials contributes to sustainability and reduces environmental impact by recycling and repurposing waste materials [11].

Coconut fibre, a bio-natural fibre derived from coconut husks, is a lightweight, eco-friendly, and cost-effective option for car door panels. It reduces weight, improves fuel efficiency, and reduces road noise. Its sound-absorbing properties also enhance acoustic comfort. Coconut fibre is a renewable resource, making it a sustainable choice for car interior materials. Its biodegradability reduces environmental impact at the end of the product's life cycle. This sustainable, quieter, and cost-effective solution meets modern automobile design needs [12, 13].

This research aims to address these challenges by systematically evaluating the noise absorption and mechanical properties of hybrid composites, optimizing their composition, and providing a foundation for sustainable and efficient car door panel material development in the automotive industry.

4.2 Methodology

4.2.1 Material

Polypropylene is used as the matrix phase, supplied by Polypropylene Malaysia Sdn. Bhd. in pellets. Rubber waste tyres (supplied by R&Z Auto and Tyre) are prepared by cleaning and cutting into rough particles, then dried and blended to obtain fine particles less than 0.5 mm in diameter. Coconut fibre is supplied by Avasia Agro Sdn. Bhd in husk form, which must be cleaned and cut into short fibres before extraction.

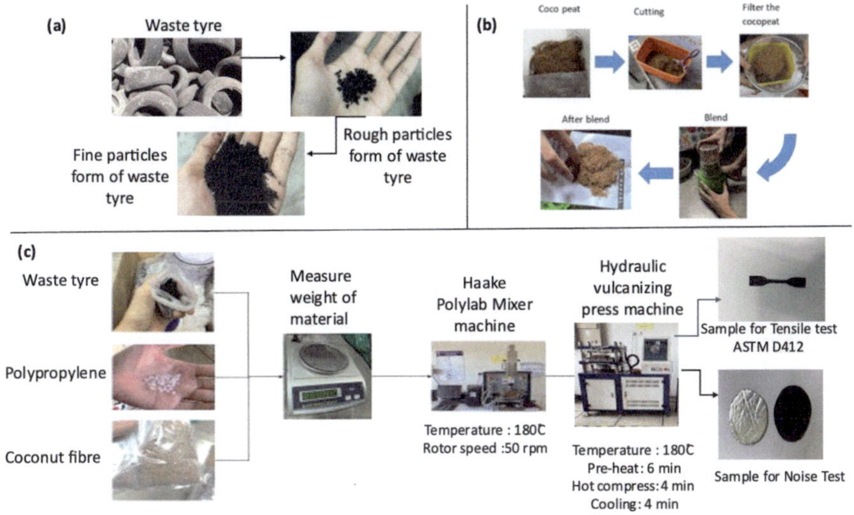

Fig. 4.1 **Xxx a** Preparation of coconut fibre; **b** Preparation of waste tyre; **c** Sample preparation for tensile and noise testing of PP/WT/CF composites

After that, it was sieved to get fine fibres around 1 mm in length. All steps of the specimen preparation are shown in Fig. 4.1.

4.2.2 *Mixing Process*

At a temperature of 180 °C and a rotor speed of 50 rpm, composites were mixed in the Haake Polylab QC Mixer Machine from Thermo Fisher Scientific Inc., United State (see Fig. 1c). Polypropylene was discharged into the mixing chamber and melted after 3 min. The waste tyre was discharged for 2 min, and fibre was inserted. The mixing process continued for 5 min until 10 min were reached for each mix. Table 4.1 shows the composite mixtures prepared with different loadings of CF fibre.

Table 4.1 Formulation of hybrid PP/WT/CF composites at different fibre loadings

Material	Composition (wt. %)
Polypropylene (PP)	70
Waste tyre (WT)	30
Coconut fibre (CF)	0, 5, 10, 15, 20

4.2.3 Hot-Press Process

The composite mixtures are then compressed in a hydraulic hot press machine (GT 7014) Dongguan Lixian Instrument Scientific Co., Ltd. Guangdong, China, to produce a ± 3.0-mm sheet of composite using the moulding of a dumbbell shape to comply with the ASTM D638 standard (see Fig. 1c). The machine was set at 10 kN forces and 180 °C. Pre-heated the mould with the sample for 6 min, hot-pressed it for 4 min, cooled it for 4 min, and then removed it from the mould.

4.2.4 Impendence Tube Testing

The impendence tube sound test measures frequency, vibration, absorption, and transmission loss in a 9-mm-diameter and 3-mm-thickness sample. The frequency ranges from 500–700 Hz, similar to a car's frequency of 500–1000 Hz [14, 15]. The test compares market-available materials for car door applications with new materials (PP/WT/CF).

4.2.5 Tensile Testing

Tensile test measures a material's ability to withstand pulling pressures and stretch before breaking. It is conducted according to ASTM D412 using a Galdabini Quasar 100 machine at 25 °C ± 3 °C, with a crosshead speed of 50 mm/min and gauge length of 50 mm. Five samples were tested for each composition.

4.3 Result and Discussion

4.3.1 Sound Transmission Loss Testing

The impendence tube test measured amplitude and frequency [16–18]. Then, the sound transmission loss coefficient was calculated using the following equation.

$$\mathrm{TL} = 20\log(\mathrm{Pa_{in}}/\mathrm{Pa_{out}}) \tag{4.1}$$

$$TL = 10\log(\tau) \tag{4.2}$$

where TL is the transmission loss and τ is the coefficient of sound transmission loss.

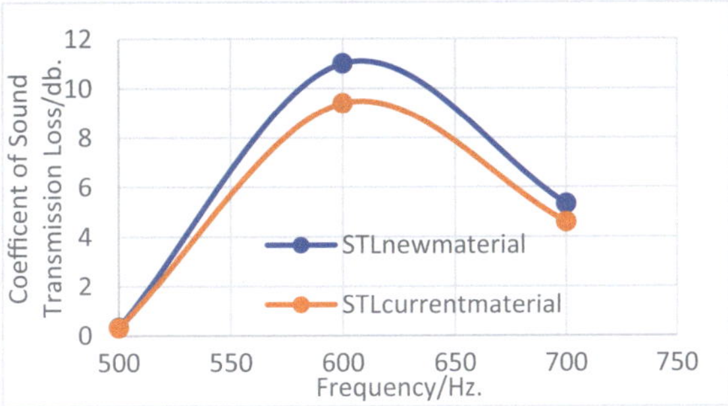

Fig. 4.2 Sound transmission loss versus frequency graph

Table 4.2 Sound transmission loss

Frequency (Hz.)	Model (Material)	Sound transmission loss/ db					
		Test 1	Test 2	Test 3	Test 4	Test 5	Average
500	New	0.32	0.36	0.31	0.31	0.36	0.34
	current	0.44	0.23	0.60	0.07	0.23	0.31
600	New	11.42	11.42	11.46	9.26	11.46	11.01
	current	10.67	10.45	9.66	9.34	6.67	9.39
700	New	0.27	12.59	0.88	12.61	0.30	5.33
	current	4.61	4.62	4.61	4.42	4.69	4.59

Figure 4.2 shows the sound transmission loss curve of new materials (PP/WT/CF composites) versus the current material used for car door panels. Each frequency and material had five measurements. Reading the coefficient of sound transmission loss from the graphs at each frequency was tabulated in Table 4.2. It shows that these two materials have identical trend lines. The new material with 5 wt. %. of CF shows higher sound transmission loss compared to current materials. In the context of sound transmission loss, more is better [19]. Sound transmission loss (STL) is a measure of a material's ability to block or reduce the transmission of sound from one side of the material to the other. It quantifies how effective a material is at isolating or attenuating sound. A higher sound transmission loss means that the material is better at preventing sound from passing through it [19]. This is particularly important in various applications, especially in automotive car door panel design, where noise control is crucial [4]

4.3.2 *Tensile Properties*

The summary of tensile properties, such as tensile strength, Young's modulus, and elongation, at break is shown in Table 4.3. Figure 4.3 shows the tensile strength for PP/WT/CF composites. It shows the optimum tensile strength at 5 wt.% of CF is

5.25 MPa. The minimum tensile strength is 2.68 MPa at PP/WT/CF with 20 wt% of CF. The tensile strength decreased with increased CF loading. The size of the filler with poor interfacial and weak bonding between hydrophilic filler and hydrophobic matrix polymer can cause a decrease [20–22]. Tensile strength decreased 50%.

Young's modulus graph trendline increases slightly from as increasing of CF loading as shown in Fig. 4.4. Composites with 20 wt.% of CF correspond to the highest modulus of elasticity of about 102.26 MPa. Increased filler loading increased the composite ductility. CF was intended to reduce composite ductility and increase stiffness. Additional more CF increases Young's modulus by about 62%. The presence of fibres in PP/WT enhanced the elastic modulus significantly [22–24].

Figure 4.5 indicates that the elongation at break is decreasing from filler loading as increasing CF loading. The maximum break elongation is at 22% and the minimum is 8.29% which occurred at 0 wt.% and 20 wt.%, respectively. The CF loading affected break elongation. When filler loading increased, composite ductility decreased significantly. Fillers made composites brittle and less stretchable [22, 25].

Table 4.3 Summary of tensile properties of PP/WT/CF composites

Composites PP/WT/CF:	Tensile strength (MPa)	Standard deviation	Young's modulus (MPa)	Standard deviation	Elongation at break (%)	Standard deviation
70/30/0	3.80	0.27	57.67	10.26	22.00	14.00
70/30/5	5.25	1.17	72.14	41.73	8.51	4.76
70/30/10	3.99	0.74	55.28	4.40	13.97	1.13
70/30/15	4.06	0.46	71.16	1.54	11.54	1.37
70/30/20	2.68	0.77	102.26	44.28	8.29	6.50

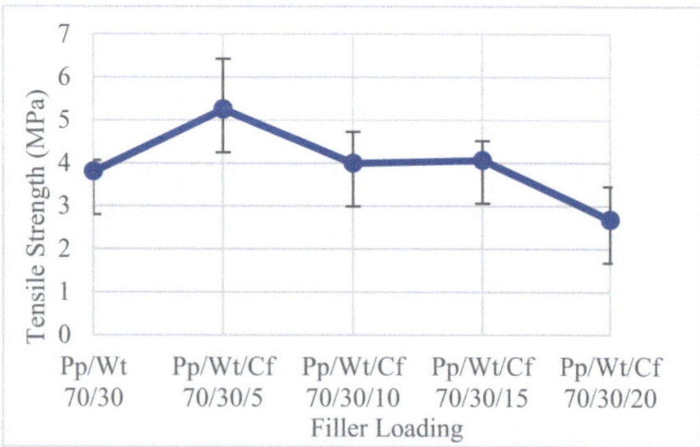

Fig. 4.3 Effect of filler loading on tensile strength at different filler loading

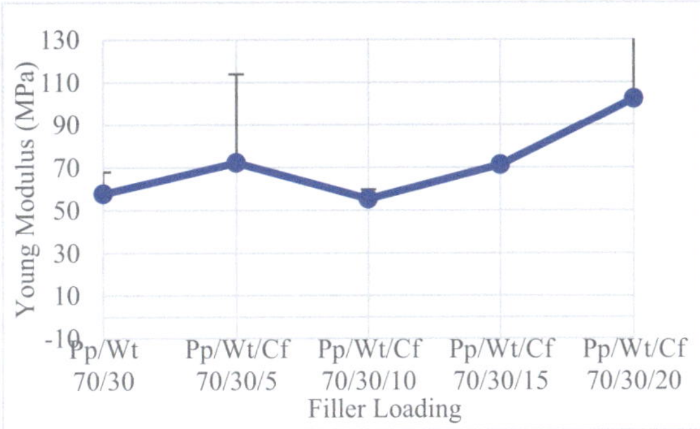

Fig. 4.4 Effect of filler loading on Young's modulus at different filler loading

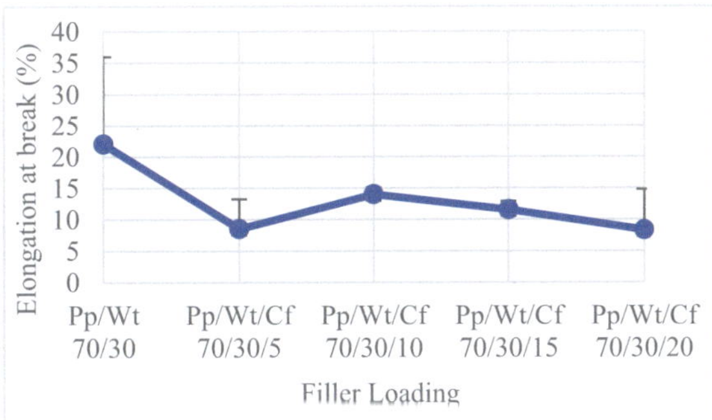

Fig. 4.5 Effect of filler loading on elongation at break at different filler loading

4.4 Conclusions

Hybrid waste rubber tyre-based polypropylene composites for car door panel applications were tested using tensile and noise absorption tests. The optimum tensile strength was 5.25 MPa with 5 wt.% of CF loading. Young's modulus increased by 62% but reduced elongation at break by 14.71%. Noise and vibration resistance tests of this composite showed higher sound transmission loss than current materials. Therefore, it is suitable for use in automotive components, such as car door panels and other quieter interior environment components.

References

1. Chen Y, Li D, Xu D (2019) Research on application of acoustic materials in automobile noise reduction. In: IOP conference series: earth and environmental science, vol 233. IOP Publishing, p 052026
2. Bakker M, Beaven R, Bogner JE, Canu P, Carvalho, Teresa M, Clarke W, Costa G, Das D, Hannan M, Hjelmar O, Imhoff P, Kalbe U, Kjeldsen P, Komilis D, Leckner B, Lü F, Oonk H, Pivato A, Polettini A, Miller SASA (2020) Encyclopedia of renewable and sustainable materials. J Clean Prod 91(4)

3. Raghu M (2018) A study to explore the effects of sound vibrations on consciousness. Int J Soc Work Hum Serv Pract 6. https://doi.org/10.13189/ijrh.2018.060302
4. Saf O, Erol H, Kutlu AE (2020) An investigation of the sound transmission loss for elastomeric vehicle door seals. Appl Acoust. 165. https://doi.org/10.1016/j.apacoust.2020.107296
5. Thomason J, Rudeiros-Fernández J (2019) Evaluation of injection molded natural fiber-polypropylenes for potential in automotive applications. In: 11th Canadian—international conference on composites, pp 1–8
6. Moritomi S, Watanabe T, Kanzaki, S (2010) Polypropylene compounds for automotive applications, vol 1. Sumitomo Kagaku: R&D Report, pp 1–16
7. Moore EP, Hanser-gardner J (1997) Polypropylene handbook. In: Polymerization, characterization, properties, processing, applications. Hanser, p 69451
8. Maurya AK, Manik G (2023) Advances towards development of industrially relevant short natural fiber reinforced and hybridized polypropylene composites for various industrial applications: a review. J Polym Res 30(1)
9. Priyadarsini M, Biswal T, Acharya SK (2023) Study of mechanical properties of reinforced polypropylene (PP)/nettle fibers biocomposites and its application in automobile industry. Mater Today Proc 74. https://doi.org/10.1016/j.matpr.2022.11.349
10. Thai QB, Chong RO, Nguyen PTT, Le DK, Le PK, Phan-Thien N, Duong HM (2020) Recycling of waste tire fibers into advanced aerogels for thermal insulation and sound absorption applications. J Environ Chem Eng 8. https://doi.org/10.1016/j.jece.2020.104279
11. Rahmani M, Adamian A, Hosseini-Sianaki A (2021) Effect of waste ground rubber tire powder on vibrational damping behavior and static mechanical properties of polypropylene composite plates: an experimental investigation. J Mater Eng Perform 30. https://doi.org/10.1007/s11665-021-06073-9
12. Adeniyi AG, Adeoye SA, Ighalo JO (2020) Sansevieria trifasciata fibre and composites: a review of recent developments. Int Polym Process 35(4):344–354
13. Adeniyi AG, Onifade DV, Ighalo JO, Adeoye AS (2019) A review of coir fiber reinforced polymer composites. Compos Part B: Eng 176:107305
14. Chi CF, Dewi RS, Huang MH (2017) Psychophysical evaluation of auditory signals in passenger vehicles. Appl Ergon 59. https://doi.org/10.1016/j.apergo.2016.08.019
15. Soeta Y, Sakamoto Y (2018) An exploratory analysis of sound field characteristics using the impulse response in a car cabin. Environments 5. https://doi.org/10.3390/environments5040044
16. Labašová E, Ďuriš R (2019) Measurement of the acoustic absorption coefficient by impedance tube. Res Pap Fac Mater Sci Technol Slovak Univ Technol 27. https://doi.org/10.2478/rput-2019-0031
17. Tang X, Yan X (2017) Acoustic energy absorption properties of fibrous materials: a review. Compos Part A: Appl Sci Manuf 101:360
18. Peng L, Lei L, Liu Y, Du L (2021) Improved mechanical and sound absorption properties of open cell silicone rubber foam with nacl as the pore-forming agent. Materials 14. https://doi.org/10.3390/ma14010195
19. Zhao Y, Xu J, Davy JL, Liu Z, Fard M (2022) Prediction of random incidence sound absorption coefficients of porous materials. Appl Acoust 189. https://doi.org/10.1016/j.apacoust.2021.108625
20. Aridi NAM, Sapuan SM, Zainudin ES, AL-Oqla FM (2016) Mechanical and morphological properties of injection-molded rice husk polypropylene composites. Int J Polym Anal Charact 21. https://doi.org/10.1080/1023666X.2016.1148316
21. Wu Q, Liu C, Li S, Yan Y, Yu S, Huang L (2023) Micronized cellulose particles from mechanical treatment and their performance on reinforcing polypropylene composite. Cellulose 30. https://doi.org/10.1007/s10570-022-04905-y
22. Ahmad SRB, Aziz MIABA, Alzahari MRB (2023) Polypropylene composites reinforced by natural and recycled fibers for bumper application: effects of fiber loading on the tensile properties. In: Progress in engineering technology V. Springer Nature Switzerland, Cham, pp 1–7, 2023

23. Várdai R, Lummerstorfer T, Pretschuh C, Jerabek M, Gahleitner M, Faludi G, Móczó J, Pukánszky B (2020) Comparative study of fiber reinforced PP composites: Effect of fiber type, coupling and failure mechanisms. Compos Part A Appl Sci Manuf 133:105895. https://doi.org/10.1016/j.compositesa.2020.105895
24. Várdai R, Lummerstorfer T, Pretschuh C, Jerabek M, Gahleitner M, Faludi G, Móczó J, Pukánszky B (2020) Reinforcement of PP with polymer fibers: effect of matrix characteristics, fiber type and interfacial adhesion. Polymer (Guildf) 190:122203. https://doi.org/10.1016/j.polymer.2020.122203
25. Adnan MAI, Zamri FA, Mohamad Sidik MS, Ahmad SR (2020) Effect of maleic anhydride polypropylene on the properties of spear grass fiber in reinforced polypropylene and ethylene propylene diene monomer composites. In: Progress in engineering technology II. Springer, Cham, pp 205–212, 2020

Chapter 5
Application of a Smart Magnetohydrodynamic Flow Sensor for Flood Prevention

Muhammad Firdaus Yahaya, Ahmad Kamal Ismail, Khairul Akmal Shamsuddin, Mior Firdaus Mior Abdul Majid, Nor Haslina Ibrahim, Mohd Fauzi Abu Hassan, and Ayub Ahmed Janvekar

Abstract Magnetohydrodynamics (MHD) is usually used in energy harvesting. Application of MHD power generators can be seen at sea, chimneys, and industry for large-scale energy harvesting. MHD produces an electromotive force by using a fluid conductor, such as plasma gas, liquid metal, or saltwater, to generate a magnetic field (e.m.f.). The electromagnetic field will be captured using an electrode made of aluminium, copper, or another conducting metal. For this project, the MHD system has been seen to have the potential to become a sensor for detecting water flow inside underground drains. This MHD sensor could help to prevent floods from occurring. Many factors has been found and most commonly that the flood incident occurred due to the overflow of rivers and clogged underground drains. Therefore, flood prevention measures must be taken immediately. This work contains a study of the MHD

M. F. Yahaya · A. K. Ismail (✉) · K. A. Shamsuddin · M. F. M. A. Majid · N. H. Ibrahim · M. F. A. Hassan
Universiti Kuala Lumpur Malaysian Spanish Institute, Kulim Hi-Tech Park, 09000 Kulim, Kedah, Malaysia
e-mail: ahmadkamal@unikl.edu.my

M. F. Yahaya
e-mail: firdaus.yahaya01@s.unikl.edu.my

K. A. Shamsuddin
e-mail: khairulakmal@unikl.edu.my

M. F. M. A. Majid
e-mail: miorfirdaus@unikl.edu.my

N. H. Ibrahim
e-mail: norhaslina@unikl.edu.my

M. F. A. Hassan
e-mail: mohdfauzi@unikl.edu.my

A. A. Janvekar
School of Mechanical Engineering, Vellore Institute of Technology, Chennai, Tamil Nadu 600127, India
e-mail: ayubahmed.janvekar@vit.ac.in

T. A. A. Razak et al. (eds.), *Progress in Engineering Technology VI*, Advanced Structured Materials 229, https://doi.org/10.1007/978-3-031-81517-1_5

system and MHD generator with different number of magnets and output power as a signal. It can be found that the more magnet used give a better signal in output power as well as a sign of a good drainage flow in underground drains.

Keywords Magnetohydrodynamics (MHD) · Electromotive force (emf) · Flood prevention

Flooding is the most catastrophic natural disaster in Malaysia and affects nearly 4.82 million people, or approximately 22% of the country's total population. There are 189 river basins in Malaysia, including Sabah and Sarawak, with the main channels going straight to the South China Sea, and eighty-five of them are prone to recurring flooding. The estimated flood disaster area is approximately 29,800 km^2, or 9% of Malaysia's total land, and it affects almost 4.3 million people [1]. The location of Malaysia consists of West Malaysia (Peninsula Malaysia) and East Malaysia (Sabah and Sarawak), and they are divided by the South China Sea. The risk and vulnerability of urban people to floods have lately grown due to an increase in impermeable surfaces, such as roads, buildings, and parking spaces.

Magnetohydrodynamics (MHD) power generators generate current without the use of mechanical parts, and the real conductor is substituted by a magnetofluid, such as plasma gas, liquid metal, or salt water [2]. Ionised fluids (gas, liquid, or plasma) serve as electrical conductors in a MHD generator. An electromotive force (e.m.f.) is formed when an ionised fluid is compelled to flow at a specific velocity through a high magnetic field. Using Fleming's right-hand rule, the direction of the generated e.m.f can be determined. It can be used to gather electric energy by crossing two electrodes over a fluid stream. A MHD power generation can be accomplished utilising either gas-plasma or liquid metal flow.

The temperature of the gas is raised throughout the thermal ionisation process to the point where electrons are ripped away from the atoms by applying more energy and can flow freely alongside the ionised atoms. MHD generators also use multi-electrode setups. Electrodes in MHD systems are primarily metal components; however, other materials can also be used. MHD power generators do not use any mechanical components to generate current, and the real conductor is substituted by a magnetofluid [3, 4]. A magnetic field can induce currents in the moving conductive fluid, creating force on the fluid, and changing the magnetic field itself [5]. If this technology is encouraged to replace all generators that produce electrical energy using gasoline, natural gas, coal, or nuclear fuel, it will help to reduce the greenhouse effect and save our planet from global warming. The greater the number of magnets and the strength of the magnetic field, the greater the current produced. The flow rate of salt water is also a crucial element because the flow of saltwater solution is required to reduce the magnetic flux. As a result, using a MHD generator to generate power for citizen usage is a wonderful concept. The MHD generator is shaped like a rocket engine and is surrounded by a large magnet. It has no moving parts and substitutes the actual conductors with ionised gas (plasma). Faraday's law of electromagnetic induction governs the MHD power generation process. There are numerous benefits to using a MHD system, for example, it eliminates thermal contamination of water. MHD has a wide range of applications for improving fuel efficiency and contributes significantly to the mitigation of serious air and thermal pollution caused by steam plants. With breakthroughs in corrosion science and superconducting magnets, rapid commercialization may be conceivable. Water's resistance varies with salinity, so each experiment used 4% (roughly seawater salinity), 7%, and 10% salinity by mass to investigate the effect of lower resistivity on the drive. Electrical power to the salt water between the electrodes could be measured using a simple circuit. Velocities at various levels of fluid pressure could be determined by analysing the distance travelled over time.

In this study, the MHD which is usually used in water flow for energy harvesting has been used as a potential sensor for detecting water flow in underground drains. If the sensor gives no signal, it can be assumed that there is no flow detection of water in the underground drain, which means that the drain is clogged, and the cleaning process must be conducted immediately.

5.1 Methodology

There are two experiments conducted; the first experiment is to measure the voltage produced by the smart magnetohydrodynamics sensor. The second experiment is to create the water flow signal. Table 5.1 shows the parameter setup of the analysis. The analysis has been divided into four setups with different number of magnets varying from 10 to 40 magnets. Salt is used as a main agent for the ionised fluid in the MHD. In Fig. 5.1, the light emitting diode (LED) is used as sensor alarm to highlight the existing of water flow in the underground drain.

Table 5.1 Analysis 1–4 setup

Apparatus	Quantity
Number of magnets	10/20/30/40 magnets
Amount of salt	0.5 kg (Cycle 1) 1 kg (Cycle 2) 1.5 kg (Cycle 3) 2 kg (Cycle 4)
Water	10 L

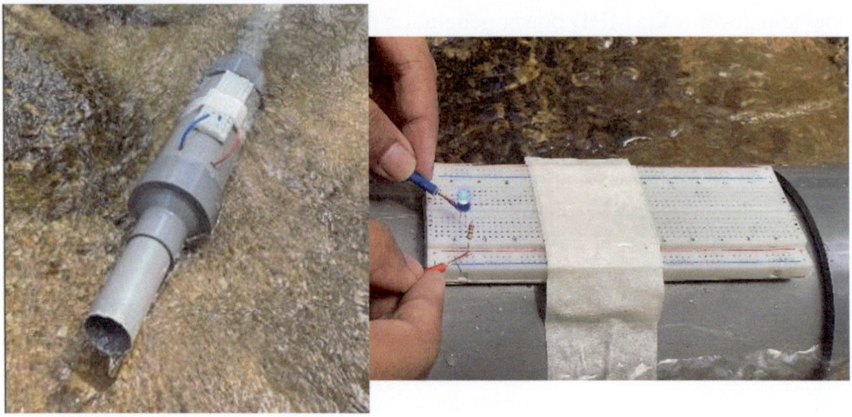

Fig. 5.1 Experimental Setup (left) with LED light up when receiving power from smart MHD sensor (right)

5.2 Result and Discussion

5.2.1 Experiment 1: Measuring the Voltage Produce Smart MHD Sensor

For analysis 1, the number of magnets used is 10 magnets. By using the result of the analysis, the data is plotted in graph shown in Fig. 5.2. For analysis 1, the result shows a linear pattern of the voltage produce by using 4 different amounts of salt and 10 magnets. For cycle 1 (0.5 kg salt), the average voltage produce is 23.27 mV, for cycle 2 (1 kg salt), the average voltage produce is 37.70 mV, for cycle 3 (1.5 kg salt) the average voltage produce is 45.57 mV, and finally, for cycle 4 (2 kg salt) the average voltage produce is 59.73 mV.

For analysis 2, the graph is shown in Fig. 5.3. The result shows a linear pattern of the voltage produce by using 4 different amounts of salt and 20 magnets. For cycle 1 (0.5 kg salt), the average voltage produce is 68.63 mV, for cycle 2 (1 kg salt), the average voltage produce is 107.20 mV, for cycle 3 (1.5 kg salt) the average voltage produce is 144.30 mV, and finally, for cycle 4 (2 kg salt) the average voltage produce is 179.07 mV.

For analysis 3, the graph is shown in Fig. 5.4. The result shows a linear pattern of the voltage produce by using 4 different amounts of salt and 30 magnets. For cycle 1 (0.5 kg salt), the average voltage produce is 138.37 mV, for cycle 2 (1 kg salt), the average voltage produce is 188.47 mV, for cycle 3 (1.5 kg salt) the average voltage produce is 232.63 mV, and finally, for cycle 4 (2 kg salt) the average voltage produce is 283.93 mV.

Finally, analysis 4 is the last analysis performed in this study. For analysis 4 as in Fig. 5.5, the result shows a linear pattern of the voltage produce by using 4 different

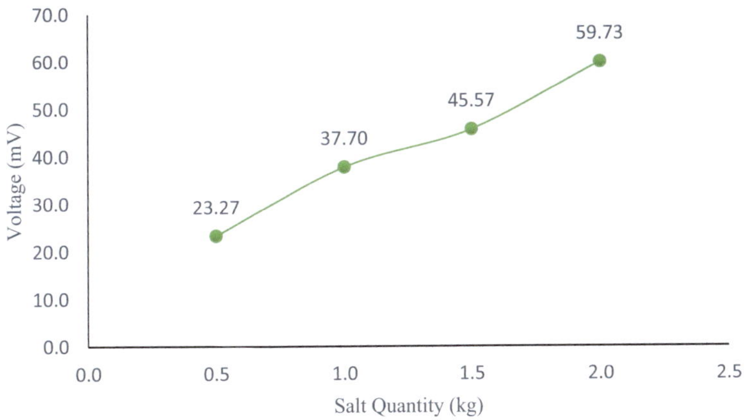

Fig. 5.2 Graph of 10 magnets voltage produces

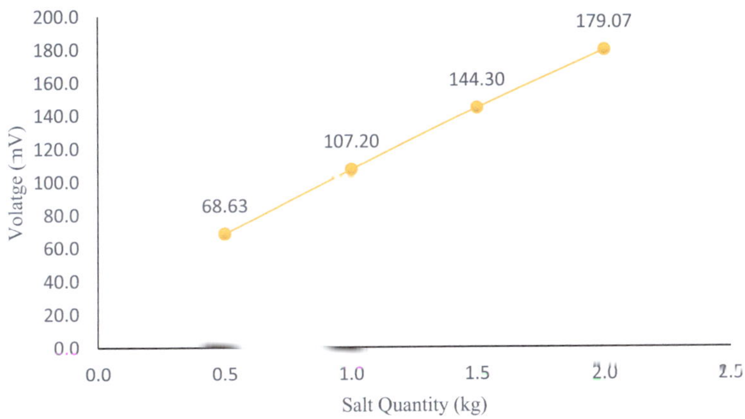

Fig. 5.3 Graph of 20 magnets voltage produces

amounts of salt and 40 magnets. For cycle1 (0.5 kg salt), the average voltage produce is 322.30 mV, for cycle2 (1 kg salt), the average voltage produce is 354.07 mV, for cycle 3 (1.5 kg salt) the average voltage produce is 373.50 mV, and finally, for cycle 4 (2 kg salt) the average voltage produce is 405.23 mV.

After completing the test, let the conductive fluid (salt water) to flow through the pipe for a few minutes. The LED is light up, indicating that it can detect water flow inside the pipe. As a result, it is possible to deduce that the magnetohydrodynamic (MHD) system can be utilised as a sensor to detect water flow inside the drainage pipe. Based on experiment, a similar linear pattern can be detected throughout the analyses performed to quantify the voltage produced by the smart MHD sensor. Several significant factors contribute to the efficiency of the smart MHD. The first is the quantity of magnets. The greater the number of magnets, the greater the magnet

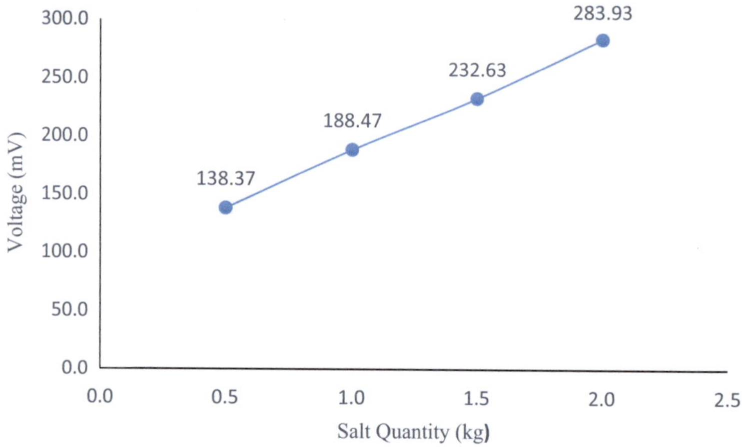

Fig. 5.4 Graph 30 magnets voltage produces

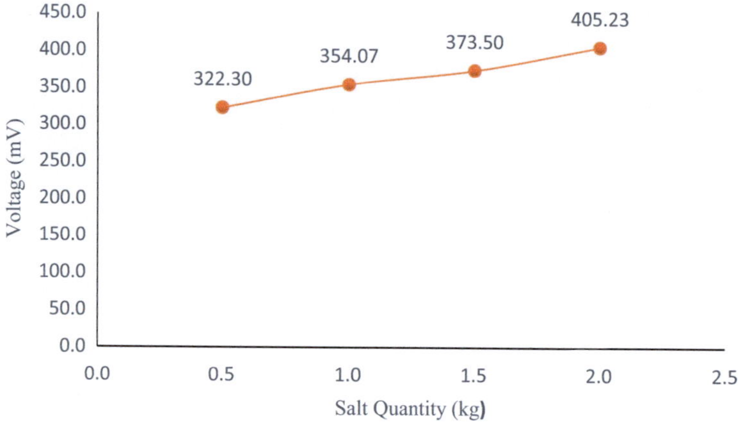

Fig. 5.5 Graph of 40 magnets voltage produces

strength and surface area of the magnetic field to produce the electromotive force that tends towards voltage. The amount of salt used in the water for the conductive fluid is the second critical factor. For experiment 1, four different quantities of salt are used: 0.5, 1, 1.5, and 2 kg. The salt was dissolved with 10 L of fresh water. As a result, the higher the salt concentration inside the conductive fluid, the higher the voltage produced. With a higher concentration of salt, there is a lot more electromotive force inside conductive fluid.

5.3 Conclusion

It can be concluded that a significantly increase of the voltage only by changing the number of magnets from 10 to 20, 30, and 40 magnets. Based on experiment, the intensity of the salt may also give a greater power output, hence giving a better water flow signal. The best amount of salt for the smart MHD sensor to detect water flow is 2 kg mixed with 10 L of water. From the result of experiment, by using 40 magnets and 2 kg of salt, it is able to produce over 400 mV, enough to power a simple LED circuit. This result proves that the MHD system can be used as a sensor. By developing the water flow signal, it can help engineer to monitor the water flow inside the underground pipe. If the LED is light up, it means that the water flow inside the underground pipe is detected and vice versa. This can give awareness to engineer to do a maintenance towards the underground pipe so that flood problem can be prevented.

References

1. Mohd Ekhwan T, Muhammad Barzani G (2014) Floods in Malaysia historical reviews, causes, effects and mitigations approach. Int J Interdiscip Res Innov 2.4:59–65
2. Dhareppagol VD, Saurav A (2013) The future power generation with MHD generators magneto hydrodynamic generation. Int J Adv Electr Electron Eng 2:2278–8948
3. Kayukawa N (2004) Open-cycle magnetohydrodynamic electrical power generation: a review and future perspectives. Prog Energy Combust Sci 30(1):33–60
4. Majid MFMA et al (2016) Fundamental studies on development of MHD (Magnetohydrodynamic) generator implement on wave energy harvesting. In: IOP conference series: materials science and engineering, vol 114(1). IOP Publishing, p 012145
5. Trapanese M et al (2018) A magnetohydrodynamic generator for marine energy harvesting. In: OCEANS 2018 MTS/IEEE Charleston. IEEE, pp 1–4

Chapter 6
Design and Development of an Innovative Walking Aid for Disabilities

Muhammad Naim Hakim Ropidon, Ahmad Kamal Ismail, Khairul Akmal Shamsuddin, Khairul Shahril Shaffee, Mohd Faizal Abu Talib, Nor Haslina Ibrahim, and Ayub Ahmed Janvekar

Abstract The ability to walk among people with disabilities and frail persons makes it difficult for them to engage in certain activities. Frail persons are classified as elderly people who have difficulties in balancing their bodies and have muscle weakness due to aging factor. Therefore, a walking aid is needed to help them to walk, improving the mobility and rehabilitation. The commercial walking aid has been developed to assist a person with disability and frail persons. However, this walking aid can be much improved. In this work, a new design has been proposed. The use of anthropometric data of elderly people has been taken into consideration in the design process. As for the structure, the critical parts of the walking aid have been examined using finite element analysis. To ensure that the new design is capable to be used, fabrication has been done according to the new design and a complete walking aid named "Guide Walker" has been produced and tested. As a result, the walking aid has a better performance compared with the existing product according to the ergonomic criteria. For the simulation result of some critical parts, the result shows that for the

M. N. H. Ropidon · A. K. Ismail (✉) · K. A. Shamsuddin · K. S. Shaffee · M. F. A. Talib · N. H. Ibrahim
Kulim Hi-Tech Park, Universiti Kuala Lumpur, Malaysian Spanish Institute, 09000 Kulim, Kedah, Malaysia
e-mail: ahmadkamal@unikl.edu.my

K. A. Shamsuddin
e-mail: khairulakmal@unikl.edu.my

K. S. Shaffee
e-mail: khairuls@unikl.edu.my

M. F. A. Talib
e-mail: mfaizal@unikl.edu.my

N. H. Ibrahim
e-mail: norhaslina@unikl.edu.my

A. A. Janvekar
School of Mechanical Engineering, Vellore Institute of Technology, Chennai, Tamil Nadu 600127, India
e-mail: ayubahmed.janvekar@vit.ac.in

T. A. A. Razak et al. (eds.), *Progress in Engineering Technology VI*, Advanced Structured Materials 229, https://doi.org/10.1007/978-3-031-81517-1_6

sitting part may accommodate 70% of the person's total body while for the armrest part is 10%. Finally, it can be concluded that the use of anthropometric data together with structure analysis is important in order to develop suitable specifications and ergonomically designed products for an elderly and disabled person and also helping them improving their movement.

Keywords Aging · Elderly people · Disabilities · Walking aid · Anthropometry dimension · Engineering design process · Finite element analysis · Smart walking aid

6.1 Introduction

According to national policy, walking difficulties among elderly people aged 60 and above is common in Malaysia which contributes to the loss of balancing and limit in movement. The problem faced among the elderly and person with disability due unfortunate event or disease has created mobility issues. The current walking aid available in the market is not suitable and needs a lot of energy to use it which can lead to another issue or secondary disease. Thus, in order to create the best product to assist in walking, the use of anthropometry data and design analysis will be helpful. Based on previous study, the use of mobility aids are devices that allow people with mobility issues to move around with more freedom and independence by reducing the impact of mobility limits, improving independence, and reducing the burden of care. These assistive technologies are critical for elderly people to have more mobility freedom, alleviate discomfort, and increase self-confidence and enjoyment in their daily lives. The objective on this research is to explore several mobility assistance device solutions and assess their benefits and downsides. As a result, the use of a standard walking aid needs a large amount of energy. It is ineffective for elderly people. In the outcome of this study, the researcher highlights the limitations of various mobility aids in order to improve the features for elderly people.

An analysis on the walker used by elders and physically disables person to maintain the balance when walking has been conducted [1]. The main features that will make a walker user-friendly and stable are the material and structure of the walker frame. The problem is that the researcher wants to analyze the load applied to the walker frame, which is foldable and has a fabric seat to allow them to rest while walking. The objective of this article is to improve stability and weight optimization, and the walker was analyzed by considering various metal combinations that are commonly used in the design of the walker. The method that has been used is that loads are applied to the various members of the walker frame to simulate both the walking and seating cases. Following the application of the loads, the results show equivalent stresses and total deformation in the walker frame. Following the results, the best metal combination is suggested for the walker's design.

This research starts with identifying the problem issues about elderly people and rehabilitation program regarding in walking and the use of assistive walking aid

[2, 3]. The conjunction of this research is proceeding with process design planning suitable for developing the walking aid [4]. To decide the design specification of the project, the use of the anthropometric dimension of the body dimension from elderly people aged 60 and above in order to develop the proper size of the project [5, 6]. Next, the design analysis was used to identify the part that could accommodate the load on the walking aid [7]. The finite element method was used to analyze the design of the walker based on the specific load in order to determine the critical part that could fail in earlier stages. There are also additional features of electronic sensors to recognize obstacles for collision avoidance when using the walking aid [8].

6.2 Methodology

There is only limited literature about perspectives on disabled people using walking aid focusing on Malaysian people. In order to get the information, the interview and questionnaire were conducted on the experienced person who had already joined the rehabilitation program held by the Ministry of Health Malaysia. The questionnaire is about the personal information of the interviewee, training provided in rehabilitation and opinion from the interviewee. The boundary condition is very important in conducting analysis where the focus part is to be conducted in this study based on the 3D design in Fig. 6.1. In this project, the focus part is the armrest position and sitting position. The load will be applied based on the realistic walking condition. This analysis is conduct by using the finite element method using the SolidWorks simulation software. The person's weight with 100 kg = 981 N is applied on both positions. According to the anthropometric dimension, the dimension of hip breath in sitting is 350 mm. It considers the higher values for hip breath in sitting for males.

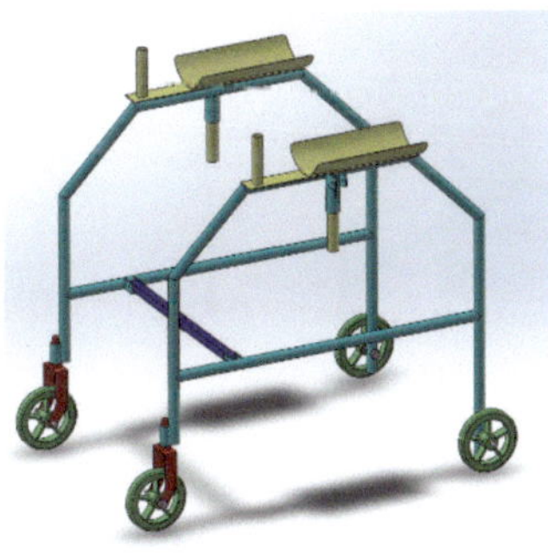

Fig. 6.1 3D model of guide walker using SolidWorks

6.3 Result and Discussion

Computational analysis

The force applied on each design equals 981 N/100 kg for the armrest position and sitting position. To consider the failure of the design, it is based on the value of yield stress which is 320 MPa. The result shows for the armrest position in Fig. 6.2 von Mises stress of 183.477 MPa, which is less than the yield stress and this means that the design is strong enough and safe. The result for the initial design of the sitting part in Fig. 6.3 indicates a von Mises stress of 602.386 MPa, which exceeds the yield stress of 320 MPa, and also tensile ultimate strength of 400 MPa. After re-designing the sitting position part 1 as shown in Figs. 6.7 and 6.8, the value of the von Mises stress reduced to 105.274 MPa and does not exceed the yield stress of 320 MPa. After that, the design improvement proceeds by adding the holding part below the sitting part, where the value of the von Mises stress is decreased to 84.825 MPa (see Fig. 6.8). It shows that the percentage improvement was made compared to an initial design with re-design part 1 of 82% and for the re-design part 1 with re-design part 2 results in an additional improvement of 19%.

Experimental testing

The comparison of the current walking aid and the new prototype guide walker has been made according to criteria of ergonomics based on body posture, safety, and comfort. In Fig. 6.6, the existing design of walking aid can be seen where the body

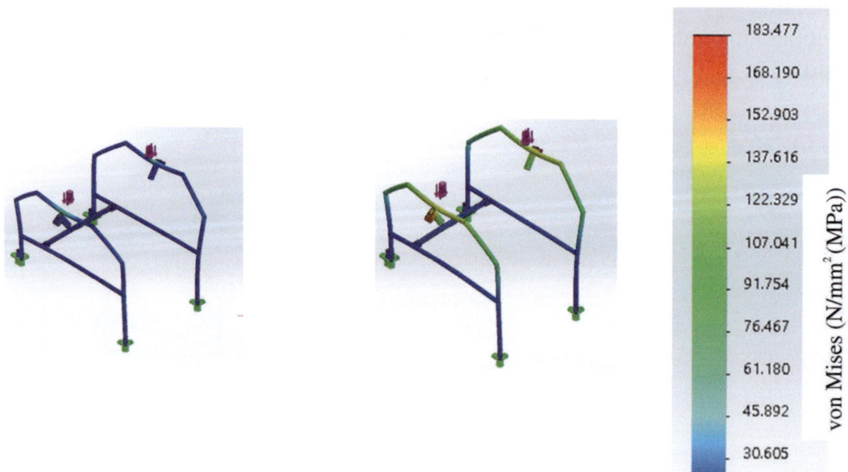

Fig. 6.2 Von Mises stress and displacement for armrest position

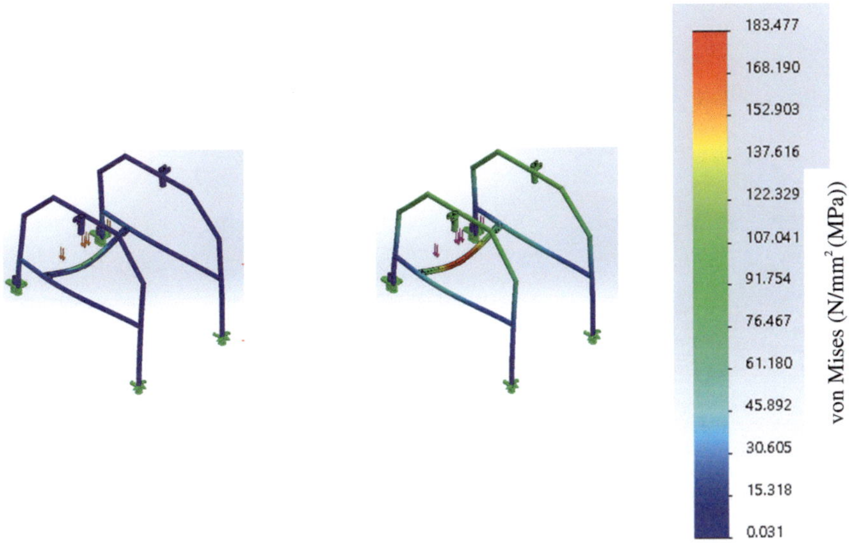

Fig. 6.3 Von Mises stress and displacement for initial design sitting position

Fig. 6.4 Von Mises stress and displacement for re-design sitting position part 1

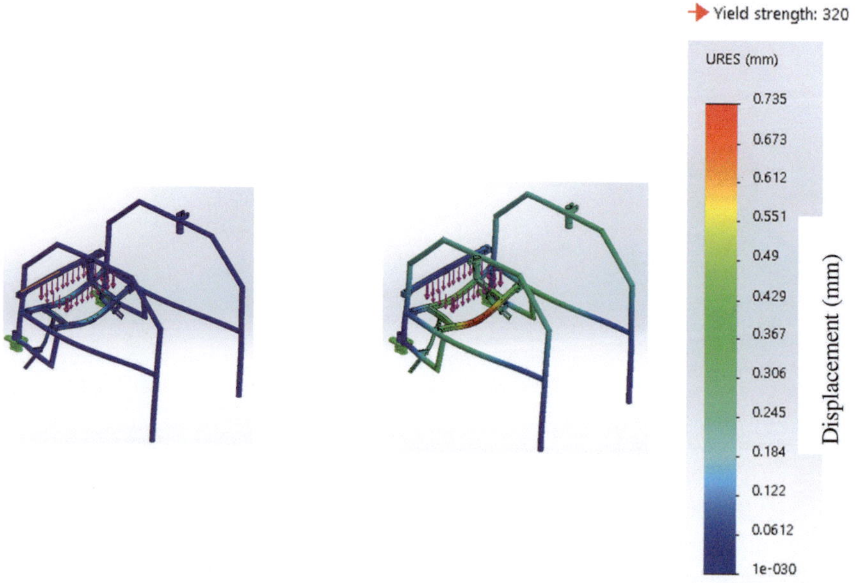

Fig. 6.5 Von Mises stress and displacement for re-design sitting position part 2

posture of the user was poor. The load was pointed on the hand grip, and there was no seat designed for the user.

While in Figs. 6.7 and 6.8, the new design of walking aid has been fabricated and tested both during standing and sitting position. It can be seen that the user was in a good body posture. The load was pointed on the forearm and the height of the

Fig. 6.6 Current design

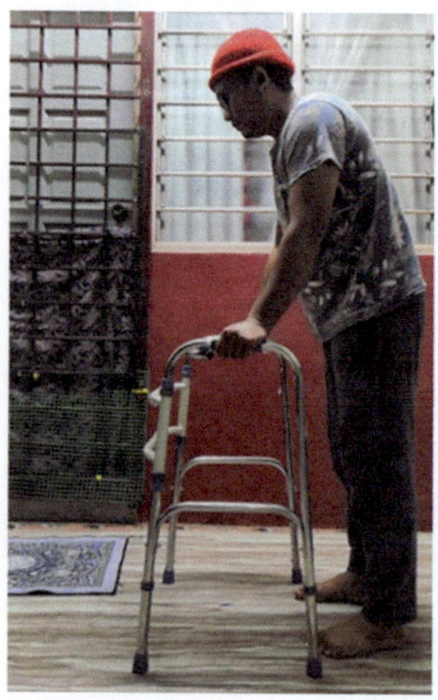

Fig. 6.7 New design

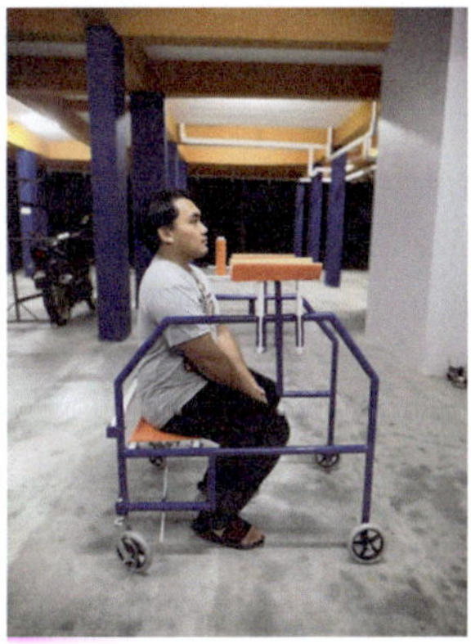

Fig. 6.8 Sitting position

armrest can be adjusted based to the user's comfort. It was also supported with soft cushion at scat, armrest, and handle for user's comfort. Wheels are attached to the walking aid and additional control system, such as brake and sensors, can be attached for safety.

The graph shown in Fig. 6.9 is based on the real testing on the new prototype guide walker. The part that received the highest load is the sitting part compared to the armrest part. The real testing at the armrest part by the participant was actually in a fixed position in order to collect the data. It could be a different result data if the real testing at the armrest part will be tested when the participant is in walking condition.

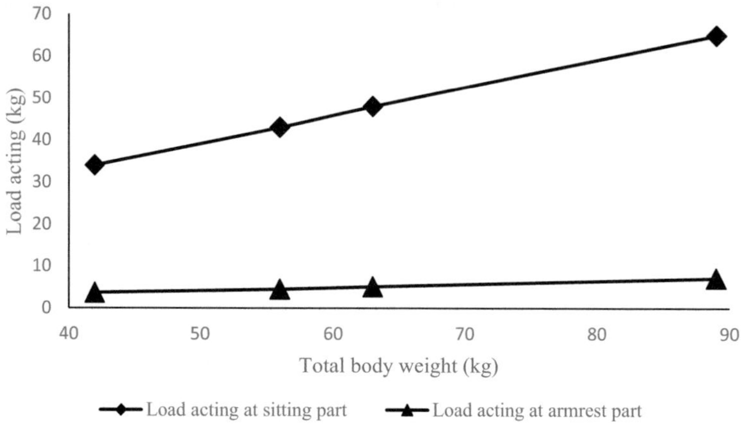

Fig. 6.9 Graph result of load acting at sitting part and armrest part which conducted in real testing of new prototype guide walker

6.4 Conclusions

In conclusion, this work has demonstrated the proper method of designing a walking aid. This project focuses on assisting elderly and disabled people to walk and increase mobility. The final design of the Guide Walker is based on anthropometric data and finite element analysis. It is important to simulate the force acting at the armrest part and sitting part for the design by comparing the maximum stress of the design and yield stress of the material. The design improvement percentage from the initial design at sitting position to re-design part 1 is 82% and re-design part 1 to re-design part 2 is 19%. The load acting on the sitting part is highest compared with the armrest part where the sitting part accommodates 70% from the total body weight compared with the armrest part is 10%.

References

1. Kulkarni A, Melavanki V, Pavate V, Abbunavar ZR (2020) Finite element analysis of a collapsible walker. Int Res J Eng Technol (IRJET) pp 2254–2260
2. Brach JS, VanSwearingen JM (2013) Interventions to improve walking in older adults. Current translational geriatrics and experimental gerontology reports, 2:230–238
3. Thies SB, Bates A, Costamagna E, Kenney L, Granat M, Webb J, Howard D, Baker R, Dawes H (2020) Are older people putting themselves at risk when using their walking frames? BMC Geriatr 20(1):1–12. https://doi.org/10.1186/s12877-020-1450-2
4. Yasin AMBM, Liong LW, Chua PS, Jianxin Z (2016) Design of an assistive walking device with special rehabilitation capabilities. Univ J Mech Eng 4(6):147–152
5. Igbokwe JO, Osueke GO, Opara UV, Ileagu MO, Ezeakaibeya KU (2019) Considerations of anthropometrics in the design of lecture hall furniture. Int J Regul Gov 7(8):374–386

6. Rosnah MY, Mohd Rizal H, Sharifah Norazizan SAR (2009) Anthropometry dimensions of older malaysians: comparison of age, gender and ethnicity. Asian Soc Sci 5(6). https://doi.org/10.5539/ass.v5n6p133
7. Bonde A, Mandavgade NK, Wankhede S, Ghate K, Nandi S, Muchulwar P, Burde P (2021) Design and analysis of walker with sit to stand assistance. Mater Today Proc 47:6074–6077
8. Chaccour K, Eid J, Darazi R, el Hassani AH, Andres E (2015) Multisensor guided walker for visually impaired elderly people. In: 2015 International conference on advances in biomedical engineering (ICABME), IEEE, pp 158–161

Chapter 7
Design and Development of a Powder-Mixed Electrical Discharge Machining Vise

Muhammad Arman Rahimi, Muhammad Al'Hapis Abdul Razak, Muhamad Husaini Abu Bakar, Mohd Shahrizan Yusoff, Mohamad Sazali Said, Mohd Riduan Ibrahim, Tajul Adli Abdul Razak, Krishnan Subramaniam, Abdul' Azeez A. Aliyu, and Ahmad Majdi Abdul-Rani

Abstract Powder-mixed electrical discharge machining (PMEDM) research is recently getting popular. The lack of suitable vise for the experiment limits the real investigation using this method. Therefore, this paper aims at creating a suitable vise to be used in PMEDM studies to optimize the electrical discharge machining (EDM) process parameters by including an electrically conductor metal powder in the

M. A. Rahimi · M. A. A. Razak (✉) · M. H. A. Bakar · M. S. Yusoff · M. S. Said · M. R. Ibrahim
Manufacturing Section, Kulim Hi-Tech Park, Universiti Kuala Lumpur Malaysian Spanish Institute, 09090 Kulim, Kedah, Malaysia
e-mail: alhapis@unikl.edu.my

M. H. A. Bakar
e-mail: muhamadhusaini@unikl.edu.my

M. S. Yusoff
e-mail: mshahrizan@unikl.edu.my

M. S. Said
e-mail: msazali@unikl.edu.my

M. R. Ibrahim
e-mail: mohdriduan@unikl.edu.my

T. A. A. Razak
Mechanical Section, Kulim Hi-Tech Park, Universiti Kuala Lumpur Malaysian Spanish Institute, 09090 Kulim, Kedah, Malaysia
e-mail: tajuladli@unikl.edu.my

K. Subramaniam
Mechanical Engineering Department, School of Engineering, Manipal International University, Putra Nilai, 71800 Nilai, Malaysia
e-mail: krishnan.subramaniam@miu.edu.my

A. A. A. Aliyu
Mechanical Engineering Department, Bayero University Kano, Kano, Nigeria
e-mail: aaaliyu.mec@buk.edu.ng

T. A. A. Razak et al. (eds.), *Progress in Engineering Technology VI*, Advanced Structured Materials 229, https://doi.org/10.1007/978-3-031-81517-1_7

dielectric fluid. In comparison with other machining methods, the expense of using EDM is greater due to the longer processing durations involved. It has been theorized that PMEDM can increase the effectiveness of EDM. PMEDM has the potential to enhance the surface smoothness of machined products, boost MRR, and reduce the tool wear (TWR). The development process of the vise includes the design using a 3D design software, design selection, machining, and measuring. Outcome from this paper shows that a new vise was developed to suit the PMEDM experiment works. For the future works, study on the powder concentration and gap voltage in SiC PMEDM is suggested. To decide how many tests need to be performed, a Taguchi orthogonal array with three levels and two factors can be utilized. The results of this study are expected to lead on the improvements in surface smoothness, TWR, machining time, and machining costs.

Keywords Powder-mixed electrical discharge machining (PMEDM) · SolidWorks · Vise

7.1 Introduction

A vise is having parallel jaws, one fixed and one moveable by screw, lever, or cam, for holding workpieces. If it will hold a workpiece during filing, hammering, or cutting, a bench vise can be permanently attached. For softer materials, metalworking vise jaws can be replaced with sheet copper or leather. Hardened steel plates with serrations grasp the workpiece. Quadruple V-shaped jaws on the pipe vise grip it in four places. With smooth jaws, woodworking vises use friction instead of serrations. To process "difficult-to-machine" materials precisely and easily, new conventional and unorthodox technologies are needed. Electrical discharge machining (EDM) is a popular non-traditional machining method. Milling electrically conductive, durable, and alloy materials that can tolerate high temperatures is widespread. These materials are used in die and mold manufacture, where EDM is common. An electrode "burns" the metal into form [1–3].

A three-dimensional electrode creates metal negative molds. Superimposing movements on the machine's major axes can change its shape, cavity, and indentation. Traditional machining processes cannot achieve this. EDM's sluggish material removal, harsh surface, and high tool wear makes it unsuitable. Powder-mixed electrical discharge machining (PMEDM) may solve these issues for EDM. PMEDM adjusts the dielectric to fulfill EDM criteria for accuracy, surface cleanliness, and material removal rate (MRR). For good discharge gap, transitivity, breakdown strength, and dielectric deionization, powder fineness is key. Scientists think that

A. M. Abdul-Rani
Mechanical Engineering Department, Universiti Teknologi Petronas, 32610 Bandar Seri Iskandar , Perak, Malaysia
e-mail: majdi@utp.edu.my

minimizing the discharge gap between the electrode and the workpiece with the right powdered material and dielectric increases EDM power consumption. EDM consistency improves roughness, MRR, and tool wear [4–6].

EDM die sinkers are widely used in machining mold inserts because of their ability to machine complex shapes with high accuracy. However, the machining is directly proportional to the length of the EDM machining time. PMEDM research is recently getting popular. The lack of a suitable vise for the experiment limits the real investigation using this method. Therefore, this paper aims at creating a suitable vise to be used in PMEDM studies to optimize the EDM process parameters by including an electrically conductor metal powder in the dielectric fluid.

7.2 Understanding Vises and 3D Design

7.2.1 Vise

A mechanical work holding device can be either manually operated or automated to hold workpieces steady for machining. In many factories, machine vises are used for jobs that need extreme accuracy. Although there are many merits associated with manual vises, machining vises are often more well suited to holding a workpiece stationary during high-speed machining operations. Because high-speed machines can cause extensive vibrations, machine vises are equipped to dampen vibrations and provide excellent stability.

Types of Vises

Face Vise—When it comes to vises, the face vise is the workhorse. It is always securely fastened to a desk, and its top is always at the same height. There are either metal or wooden canines within. To prevent metal jaws from scratching or denting the work, wooden cheeks are typically employed. Dogs that can be detached from the movable jaw and used as bench dogs are a potential addition. Quick release woodworker vises have a screw that may be engaged or disengaged with a quarter turn of the handle. If the jaw is unlocked, it can be opened as far as it goes in any direction. As a result, acclimation to new situations occurs far more rapidly. Acme and buttress threads see heavy use. The end (or tail) vice will be installed on the right side of the workbench's narrow end, and the face (or front) vice will be installed on the front, left side of the long side, for a right-handed worker.

Bench Vise—When clamping metal, an engineer's vise, sometimes called a metal-working vise, machinist's vise, or, colloquially, a "bench vise," is employed. This holds the metal while filing or sawing. Cast iron is the most common material; however, cast steel and malleable cast iron are also used. Jaws are often detachable and can be swapped out for a new set. Aluminum, copper, wood (for woodworking), or plastic jaw covers with soft padding can be utilized to safeguard fragile materials. Engineer's vises typically have jaw openings that are the same width as the jaw width, though they can be larger if necessary.

Machine Vise—Machine vises can be found installed on drill presses, grinding wheels, and mills. Abrasive chop saws feature built-in vises that are either coarse-threaded or cam-actuated machine vises.

7.2.2 3D Design

Robotics, electronics, computers, and control systems have created smartphone apps and autonomous cars. Digital modeling and development are needed before manufacturing. SolidWorks, a solid modeling CAD/CAE package, is popular among mechatronics engineers. In 1997, Dassault Systems bought SolidWorks from MIT graduate Jon Hirschtick. The suite now has 2D and 3D modeling tools. SolidWorks builds mechatronics systems. Software can aid project planning, visual ideation, modeling, feasibility evaluation, prototyping, and management. Then, the software is incorporated into the creation of hardware, software, and other important aspects. Device management, analytics, data automation, and cloud services are other programmed uses.

7.3 Methodology

This project started by sketching, designing the vise on SolidWorks conducting the design selection matrix, and fabricating the jaws. SolidWorks software was used to run the analysis. The analysis that has been done was to see any limitations for the vise. The vise design specifications are as follows.

(1) Size restrictions—The vise dimension should not exceed 100 cm × 65 cm. The base does not exceed 38 cm × 65 cm.
(2) Manufacturing Process—Major parts of the assembly were to be fabricated using tooling and equipment available in a workshop. Raw material and fasteners available on the local market must be used in the assembly wherever possible.
(3) Maintenance—Clamps should be easily removed from the vise at the end of every operation. When it is not in use, the clamp should be stored in a clean dry environment. Any parts made of steel must be regularly oiled to prevent surface corrosion.
(4) Target Costs—The total cost of raw materials and parts for manufacture should be less than the cost of a standard vise.
(5) Installation—The clamp is to be installed according to the workpiece size.

7.3.1 Designing Stage

Sketching is seen as an important way to express creativity and one of the best ways to think visually, so it is used in design. It is the main way that design engineers get their ideas out into the world. Drawings serve as visual cues for further refining and changing.

- Design A

This design is an inspiration from the CNC bench vise, see Fig. 7.1.

- Design B

This design will use the pneumatic cylinder mechanism as the clamp tools, see Fig. 7.2. The pneumatic cylinder will move by each cylinder to the desire shape of the workpiece.

- Design C

This design was the magnetic chuck that always has been used in EDM die sinker with magnets that can grip the workpiece strongly, see Fig. 7.3.

Fig. 7.1 CNC bench vise

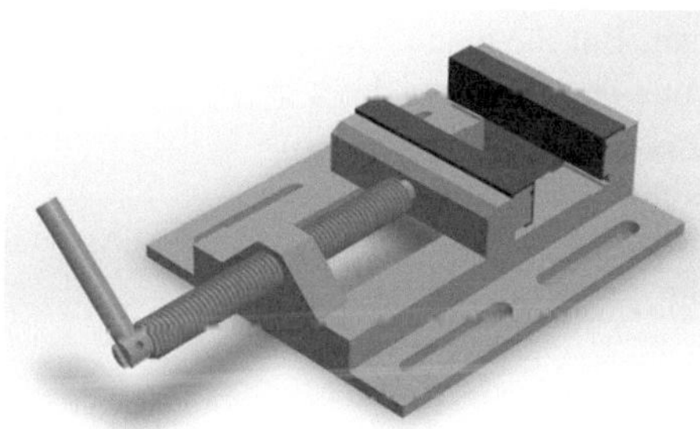

Fig. 7.2 Pneumatic cylinder

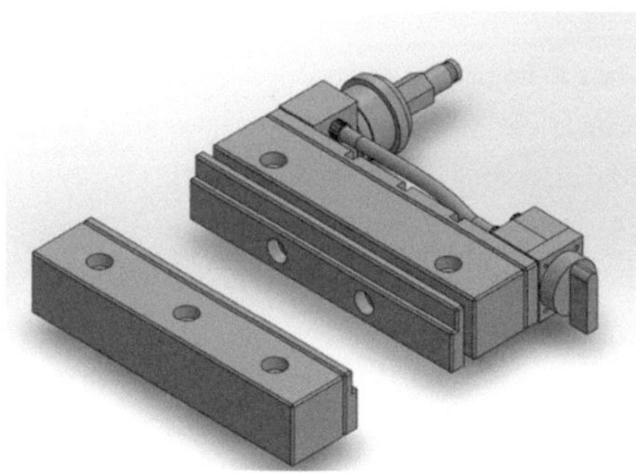

Fig. 7.3 Magnetic chuck

7.3.2 Design Selection Matrix

From Table 7.1, design A meets a criterion by getting the rating five for the first criteria because the vise can easily be fabricated. The manufacturing cost is lower than for others designs. Also, the flexibility to clamp the workpiece on the experiment also meets with the requirement. This design can be handled only by one person.

For design B, it has a good design jig, which can clamp any shape by having several numbers of pneumatic cylinders that can easily adapt to the desired shape of part. But the manufacturing cost is higher than design A. Also, the maintenance is complicated due to the pneumatic cylinder.

Lastly, for design C, this is the normal vise that is used on the EDM die sinker which is the magnetic chuck. The cost for the chuck is too high due to the mechanics of the vise. The flexibility is high because it can clamp any shape. Also, to fabricate the magnetic chuck is not easy.

Table 7.1 Design selection matrix

Criteria	Weight factor	Rating			Weight factor X rating		
		A	B	C	A	B	C
Easy to fabricate	5	5	3	3	25	15	15
Safe to use	5	4	4	4	20	20	20
Manufacturing cost	5	5	3	2	25	15	10
Flexibility	4	5	5	5	20	20	20
Reduce setup time	4	5	3	5	20	12	20
Maintenance	3	4	3	5	12	9	15
Ergonomics	3	5	4	5	15	12	15
Total					137	113	115

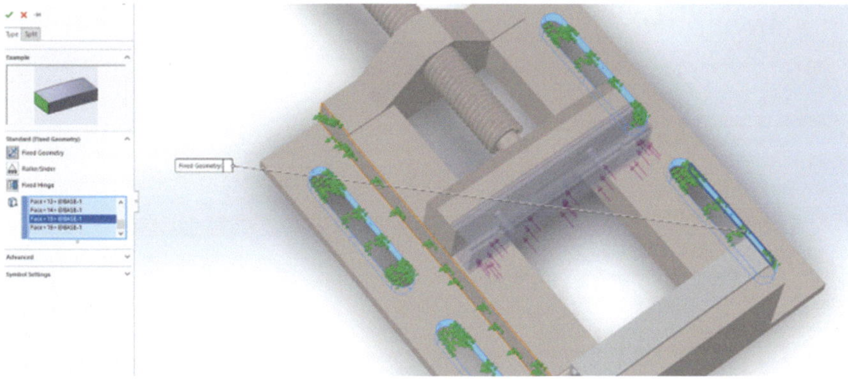

Fig. 7.4 Fixtures

7.3.3 Selected Design

From the design selection, design A was chosen based on the criteria from the rating matrix. The model was generated by using SolidWorks. The material used was an aluminum alloy. Different sizes of the jaw plate were developed: 3 mm, 5 mm, 7 mm, 9 mm, and 11 mm. The model was generated part by part using features sketch, extrude cut, etc., that features in the SolidWorks itself.

7.3.4 Finite Element Analysis

In this project, the material selected to run the finite element analysis (FEA) was aluminum. As shown in Fig. 7.4, the model was constrained at the bottom of the vise because it will stay static when a real situation occurs. The meshing technique that was used is mesh control, which is available in SolidWorks.

7.4 Result and Discussion

7.4.1 Analysis of Vise

From the analysis done, it can be said that the displacement value obtained from finite element analysis in SolidWorks is smaller than in theory. The stress value when the force was applied was 396,992 N/mm^2. The yield strength of mild steel was higher, and this will not affect the vise because it did not exceed the value of yield strength. So, no plastic deformation occurred. For the strain value, it was also small, at 0.5584, see Fig. 7.5.

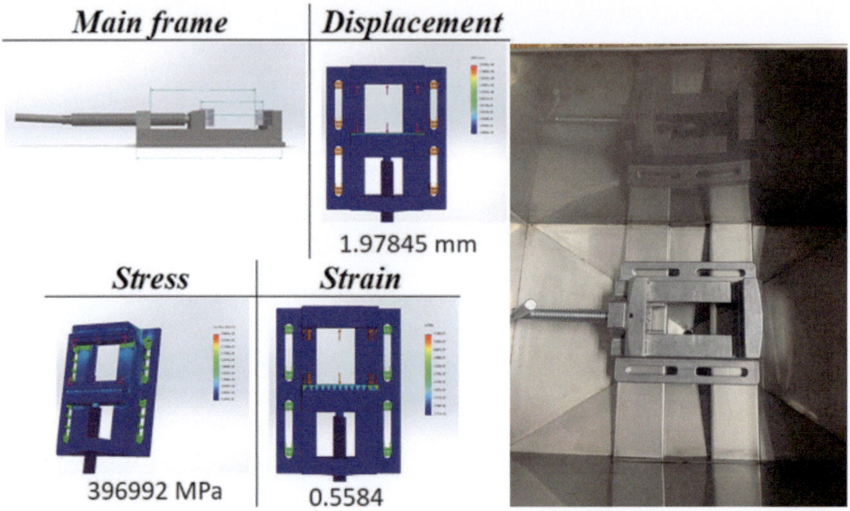

Fig. 7.5 Analysis of vise and fabricated vise located in PMEDM tank

This vise has been fabricated and been fixed into the custom tank for the PMEDM experiment. This tank will be placed in the EDM and will be filled with dielectric fluid from the outside of the tank. For the inside, the dielectric fluid will be mixed with a certain amount of silicon carbide. The vise will be placed on a platform inside the tank. Also, the jaws will be installed with the correct size of the workpiece. The jaws have reduced the setup time for finding the flatness. If the clamps are not built in step design, it is difficult to allocate 90° for each side of the workpiece to ensure that the workpiece is in the right position. There are five types of jaws that have different sizes that can be used with suitable workpiece on the PMEDM experiment as shown in Fig. 7.6.

Fig. 7.6 Fabricated 3-mm, 5-mm, 7-mm, 9-mm, and 11-mm jaw

7.5 Conclusion

The design of the project is a significant part in this project. Based on the analysis that has been done on the material selection, PMEDM and EDM vise have been a help on developing a new vise for this PMEDM research work. The design and fabrication have been done and proven successful even though may have several problems encountered with the project.

Through the analysis and simulation that have been done using SolidWorks software, it ensures that the vise will be safe with a proper limitation. Also, several characteristics of choosing the design by using design selection matrix made the process easier. Since using FEA analysis, the project will show its limitations. The fabrication of the project has been successfully done. The project has been fabricated with the desired dimension based on the design that has been conducted in SolidWorks. The clamp was successfully fixed with the vise and the tank as prediction.

In the future, here are some suggestions for this project. The first suggestion is to use other materials, such as stainless steel or aluminum since this material will avoid corrosion. Next, the magnetic base will be good if we can build it into this vise.

References

1. Luan ND, Minh ND,Thanh LTP (2019) Multi-objective optimization of PMEDM process parameter by topsis method. Int J Trend Sci Res Dev 3(4):112–115. https://doi.org/10.31142/ijtsrd23169
2. Razak MA, Rani AMA, Saad NM, Littlefair G, Aliyu AA (2018) Controlling corrosion rate of Magnesium alloy using powder mixed electrical discharge machining. IOP Conf Ser Mater Sci Eng 344(1). https://doi.org/10.1088/1757-899X/344/1/012010

3. Tran T-H, Nguyen M-C, Luu A-T, Do T-V, Le T-Q, Vu T-T, Tran N-G, Do T-T, Vu N-P (2020) Electrical discharge machining with sic powder-mixed dielectric: an effective application in the machining process of hardened 90crsi steel. Machines 8(3):36. https://doi.org/10.3390/machines8030036
4. Deepthi S, Dharmendra BV (2019) A review of current research on powder mixed electric discharge machining. Int J Mech Prod Eng Res Dev 9(3):453–462. https://doi.org/10.24247/ijmperdjun201948
5. Joshi AY, Joshi AY (2019) A systematic review on powder mixed electrical discharge machining. Heliyon 5(12) https://doi.org/10.1016/j.heliyon.2019.e02963
6. Abdudeen A, Qudeiri JEA, Kareem A, Ahammed T, Ziout A (2020) Recent advances and perceptive insights into powder-mixed dielectric fluid of EDM. Micromachines 11(8):754. https://doi.org/10.3390/MI11080754

Chapter 8
Effect of Louvered Fin Angle for an Automotive Radiator on the Flow Structure and Velocity Field

Muhamad Afif Aqil Mohd Safrurazi, Faiza M. Nasir, and Siti Lydia Rahim

Abstract The performance of automotive radiator fins greatly depends on their configuration and layout. Despite various designs, optimization of these fins, especially in terms of angle, has been limited due to their small size. Radiators play a critical role as heat exchanger devices, ensuring efficient engine operation. This study focuses on analyzing and simulating the flow patterns and velocity fields around louvered fins. Specifically, the project investigates the airflow behavior through louver fins at different angles: 20°, 25° and 30°. To test the hypothesis that varying louver angles result in distinct airflow patterns, experiments were conducted in a desktop wind tunnel. A smoke generator aided in visualizing the airflow, and a high-speed camera captured the motion in slow motion. Image analysis, employing the PIVlab toolbox in MATLAB, provided velocity field data. The results demonstrated that louvered fins induce notable flow vortices and eddies, particularly at a 25° angle, where the mean air velocity is the highest.

Keywords Wind tunnel · Radiators · Louvered fin · Heat exchangers

M. A. A. M. Safrurazi
Kulim Hi-Tech Park, Universiti Kuala Lumpur Kampus Cawangan Malaysian Spanish Institute, 09000 Kulim, Kedah, Malaysia
e-mail: mafif.safrurazi@s.unikl.edu.my

F. M. Nasir · S. L. Rahim (✉)
Centre for Women Advancement and Leadership, Universiti Kuala Lumpur, 50250 Kuala Lumpur, Kedah, Malaysia
e-mail: sitilydia@unikl.edu.my

F. M. Nasir
e-mail: faiza@unikl.edu.my

T. A. A. Razak et al. (eds.), *Progress in Engineering Technology VI*, Advanced Structured Materials 229, https://doi.org/10.1007/978-3-031-81517-1_8

8.1 Introduction

Louvered fins find extensive use as airside heat transfer surfaces in heat exchangers like automotive radiators, air-cooled oil coolers, intercoolers, and heaters. Wang et al. [1] conducted a comparative study of airside performance in fin and tube heat exchangers, concluding that the heat transfer coefficient of louvered fins surpasses that of semi-dimple vortex generators and plain fin geometries. Okbaz et al. [2] presented a study on the heat transfer and pressure drop characteristics of a louvered fin and round tube heat exchanger with different louvered fin geometries. Numerous other studies have also reported superior performance of heat exchangers employing louvered fins [3].

Many researchers have investigated the impact of geometric properties of louvered fins, such as louver angle, louver pitch, flow length, and fin pitch, on the thermal performance of heat exchangers [2–7]. Findings from these studies have highlighted that the heat transfer performance of louvered fins is significantly affected by the complex and varied flows influenced by configuration, arrangement, and geometry. Understanding and subsequently improving or optimizing louvered fins necessitate a thorough comprehension of the flow structure around the louvers.

Flow visualization can be achieved through wind tunnel experiments and computational fluid dynamics (CFD) simulations. While simulations offer detailed insights into the enhancement mechanisms of louver geometry, results may be susceptible to assumptions. On the other hand, wind tunnel experiments employ scaled-up models of louver fins, allowing for flow visualization through smoke tracing [8, 9] and dye injection [10, 11]. Limited experimental studies have been published concerning flow visualization on louvered fins, partly due to the costs and complexities associated with operating wind tunnels. In fact, more experiments should be conducted, particularly at low Reynolds numbers [2].

Hence, this study's primary objective is to conduct a flow visualization study of the airflow around louver fins with varying louver angles to comprehend the flow structure and its behavior comprehensively.

8.2 Methodology

8.2.1 Design and Development of the Wind Tunnel

The Test Section

The first step in the methodology was the design and the preparation of the test section, i.e., the louvered fins of the radiator. The louvered fins were prepared with three different angles, 20°, 25°, and 30° (Fig. 8.1). It was then placed in a desktop wind tunnel to simulate the airflow surrounding the fins. The cross-section was set at 15.x.15 cm^2. To ensure that the length-to-diameter ratio of the test section (d) exceeds

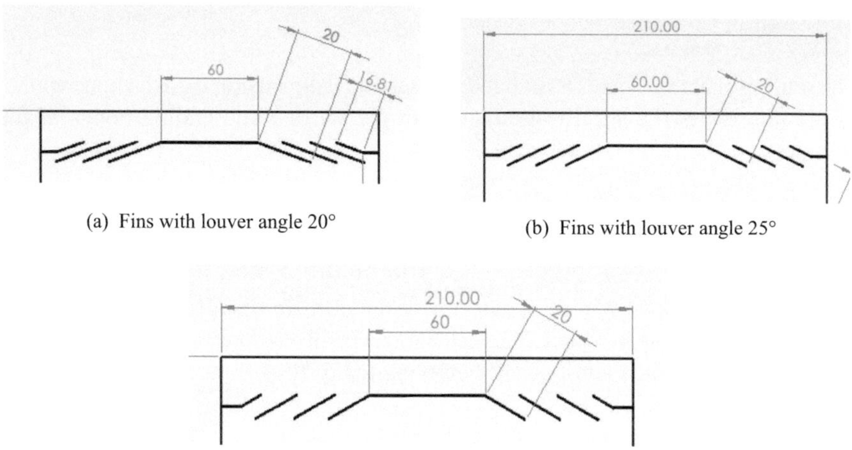

(a) Fins with louver angle 20°

(b) Fins with louver angle 25°

(c) Fins with louver angle 30°

Fig. 8.1 Schematic drawing of test section **a** Fins with louver angle 20°; **b** Fins with louver angle 25°; **c** Fins with louver angle 30°

1.5, the length was calculated accordingly. Thus, the chosen length is equivalent to the diameter (d) multiplied by 1.5, resulting in 22.5 cm.

The Contraction Cone

The contraction cone ratio in a low-speed wind tunnel typically varies between 4 and 9, depending on the cone's length and its contraction ratio concerning the test section. In this case, the contraction ratio was calculated as four based on the chosen dimensions of the cone. The cross-sectional dimensions were defined as 30 x 30 cm^2 at the outer end and 15 x 15 cm^2 at the inner end. To maintain a length-to-diameter ratio (L/D) of 1.5, the length of the cone was determined to be 45 cm, achieved by multiplying the diameter (30 cm) by 1.5.

Honeycomb

The honeycomb dimensions were tailored to align with the cross-section at both ends of the contraction cone. Consequently, the honeycomb's size was determined to be 30 × 30 cm. To meet the requirement of a minimum thickness of 1.25 inches (approximately 6.4 cm) for the honeycomb, a thickness of 3 cm was deemed appropriate based on the design, aimed at minimizing turbulence.

The Settling Chamber

The settling chamber shares a similar cross-sectional dimension with the honeycomb, as it conforms to the specifications of both the starting and ending points of the contraction cone (which is $30 \times 30\ \mathrm{cm}^2$).

The Diffuser

The diffuser is designed in accordance with the diffuser cone's specified diffusion angle of approximately 5^0 (φ). The calculation involves determining the diffusion angle ($\varphi/2 = 2.54$) and subsequently deriving the diffuser's outer diameter (D_o) to be 19 cm.

Smoke Generator

Smoke was generated using dry ice, where the sublimation process ensued from the interaction between boiling water and the dry ice. This resultant smoke was then channeled into the wind tunnel through a specially designed tube to streamline its flow.

8.2.2 *Image Capture and Image Analysis*

Image analysis was conducted to meticulously observe flow visualization within the wind tunnel. To optimize the precision of the flow visualization results, a high-speed camera was employed to capture images of the flow. This camera effectively recorded the movement of airflow through the fins in slow motion. The captured images were then compared with simulations created for the identical fin design.

PIVlab represents a particle image velocimetry (PIV) software program featuring a graphical user interface (GUI). Functioning as a MATLAB toolbox, PIVlab estimates velocities in image data with exceptional ease of use, applicable to both movies and images.

8.3 Results and Discussion

8.3.1 Flow Structure and PIV Images

Figure 8.2 shows the flow visualization and PIV images, respectively, for fin louver angle of 20° at different frames. In the smoke flow structure, no eddy can be detected at frame 1 is the starting point of the flow visualization process. At frame 70, distinct patterns of eddies and swirling flows are noticeable at the redirection louver and the middle section. These vortices illustrate the occurrence of turbulent flow within the fin. The PIV image for frame 70, showcasing velocity magnitude contours, precisely identifies high velocity magnitudes at the same regions where the eddies and vortices are observed. For frame 150, flow behavior resembles laminar behavior and it is clearly evident from the PIV image that there is no presence of eddies, indicating that the flow has become steady and uniform due to the presence of the louver angle fin.

Figure 8.3 shows the flow visualization and PIV images, respectively, for fin louver angle of 25° at different frames. At the initial frame 1, a vortex is evident beneath the redirection louver at an earlier stage. The PIV image for frame 1 featuring velocity magnitude contour demonstrates an aggregation of high velocity magnitudes under the redirection louver or within the midsection, indicating the presence of turbulence and eddies. At the 40th frame, an eddy is observed in the initial region, indicating the onset of turbulence as the smoke passes through the fin. From the PIV image, there is

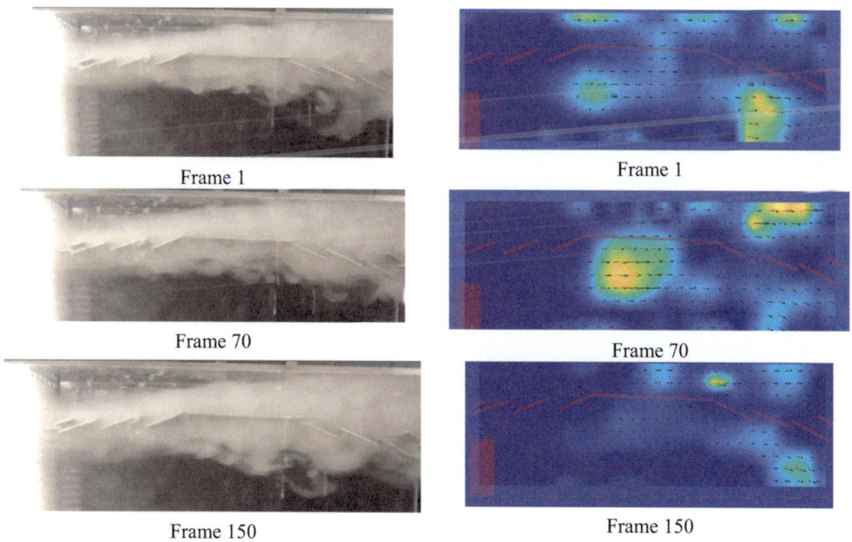

(a) Flow visualization (b) PIV image

Fig. 8.2 **a** Flow visualization; **b** PIV image for fins with louver angle of 20°

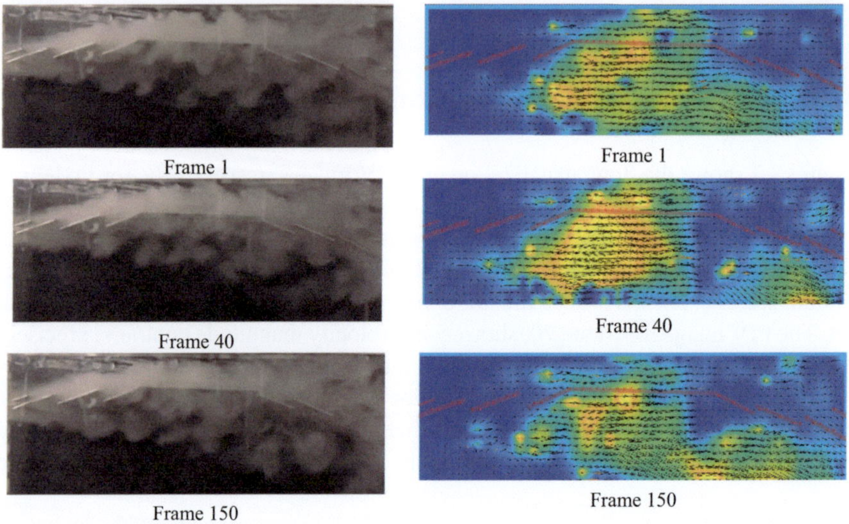

(a) Flow visualization (b) PIV image

Fig. 8.3 Flow visualization and PIV images surroundings fins with louver angle of 25°

a significant eddy, particularly in the area around the redirection louver, manifesting as yellow contours denoting high velocity magnitude.

The flow and the PIV images from all the frames reveal smoke accumulation at the starting region, implying that a 25° angle for the fin efficiently facilitates the transition from laminar to turbulent flow, featuring eddies and vortices. The results also display heightened airflow speed around the fin and in the bottom right corner of the test section, resulting from the convergence of smoke from the upper region. At this louver angle of 25°, turbulence flow can be observed around the fins, especially in the extensive high velocity area beneath the redirection louver.

Figure 8.4 shows the flow visualization and PIV images, respectively, for fin louver angle of 30° at different frames. The flow visualization for frame 1 depicts the steady flow of smoke passing through the fin structure. The smoke exhibits consistent movement, entering from the 1st fin segment and exiting at the 6th fin segment. Notably, eddies are discernible at the 3rd segment, where the flow swirls. The PIV image for frame 1 reveals contours indicating high velocity at the top of the 3rd fin segment and beneath the middle section. This is where eddies, vortex formations, and high velocity magnitude prevail. Images in frame 20 distinctly displays eddies at the 2nd, 3rd, and the redirection louver. These locations exhibit evident turbulence in the flow. PIVLAB detects high velocity magnitudes primarily under the redirection louver.

The flow structure for different angles of fins was analyzed in detail. The flow for fin angle 20° showed that the air flow was not uniform and had a high velocity near the center of the radiator. The flow for fin angle 25° showed that the air flow was more uniform than the 20° angle, but still had a high velocity near the center of the

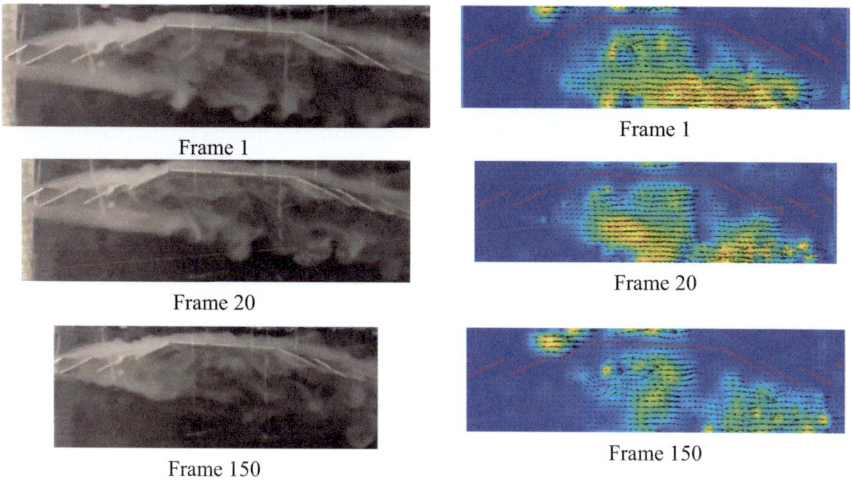

(a) Flow visualization (b) PIV image

Fig. 8.4 Flow visualization surroundings fins with louver angle of 30°

radiator. The flow for fin angle 30° showed that the air flow was the most uniform and had a lower velocity near the center of the radiator.

8.3.2 Mean Velocity

Using PIVlab, the velocity of the air flows surrounding the fins can be determined. The mean air velocity at different regions of the fins is summarized in Table 8.1.

Table 8.1 illustrates that fins set at a louver angle of 25° exhibit the highest mean velocity. This can be attributed to the presence of numerous eddies and vortices within the fin configuration. The mean velocity across the entire fin configuration at a 25° angle stands at 1.47 m/s. This value falls within the desired range of 0.9–1.9 m/s, in line with anemometer-recorded velocity values.

In summary, all average velocity values obtained fall within the acceptable and valid range of 0.9–2.0 m/s, confirming the satisfactory performance of the fin configurations.

Table 8.1 Mean air velocity at different region of louvered fins

Fin angle (°)	Starting region (m/s)	Middle region (m/s)	Trailing region (m/s)
20°	0.29675	1.2768	1.117
25°	0.83301	2.2054	1.3817
30°	0.55464	1.6933	1.0752

8.4 Conclusions

This study explored the effect of the angle of louvered fins on the flow structure and velocity field of the louvered fins on automotive radiators. It was found that the angle of the louvered fins has a significant effect on the flow structure and velocity field of the fins. The results showed that the louver angle of 25° was the most effective in terms of improving the flow structure and velocity field. This finding is important for the automotive industry as it can lead to the development of more efficient and effective radiators, which can improve the performance of vehicles.

References

1. Wang CC, Chen KY, Liaw JS, Tseng CY (2015) An experimental study of the air-side performance of fin-and-tube heat exchangers having plain, louver, and semi-dimple vortex generator configuration. Int J Heat Mass Transf 80:281–287. https://doi.org/10.1016/j.ijheatmasstransfer.2014.09.030
2. Okbaz A, Pınarbaşı A, Olcay AB, Hilmi Aksoy M (2018) An experimental, computational and flow visualization study on the air-side thermal and hydraulic performance of louvered fin and round tube heat exchangers. Int J Heat Mass Transf 121:153–169. https://doi.org/10.1016/j.ijheatmasstransfer.2017.12.127
3. Tran N, Wang CC (2020) Optimization of the airside thermal performance of mini-channel-flat-tube radiators by using composite straight-and-louvered fins. Int J Heat Mass Transf 160. https://doi.org/10.1016/j.ijheatmasstransfer.2020.120163
4. Ferrero M, Scattina A, Chiavazzo E, Carena F, Perocchio D, Roberti M, Toscano Rivalta G, Asinari P (2013) Louver finned heat exchangers for automotive sector: numerical simulations of heat transfer and flow resistance coping with industrial constraints. J Heat Transfer 135. https://doi.org/10.1115/1.4024758
5. Karthik P, Kumaresan V, Velraj R (2015) Experimental and parametric studies of a louvered fin and flat tube compact heat exchanger using computational fluid dynamics. Alex Eng J 54:905–915. https://doi.org/10.1016/j.aej.2015.08.003
6. Qi ZG, Chen JP, Chen ZJ (2007) Parametric study on the performance of a heat exchanger with corrugated louvered fins. Appl Therm Eng 27:539–544. https://doi.org/10.1016/j.applthermaleng.2006.06.015
7. Vaisi A, Esmaeilpour M, Taherian H (2011) Experimental investigation of geometry effects on the performance of a compact louvered heat exchanger. Appl Therm Eng 31:3337–3346. https://doi.org/10.1016/j.applthermaleng.2011.06.014
8. Davenport CJ (1983) Correlations for heat transfer and flow friction characteristics of louvred fin. In: AIChE symposium series. pp 19–27
9. Webb RL (1990) The Flow structure in the louvered fin heat exchanger geometry. SAE trans
10. Huisseune H, T'Joen C, De Jaeger P, Willockx A., De Paepe M (2010) Study of junction flows in louvered fin round tube heat exchangers using the dye injection technique. Exp Therm Fluid Sci 34. https://doi.org/10.1016/j.expthermflusci.2010.05.006
11. Huisseune H, T'Joen C, Jaeger P De, Ameel B, Schampheleire S De, Paepe M De (2013) Performance enhancement of a louvered fin heat exchanger by using delta winglet vortex generators. Int J Heat Mass Transf 56. https://doi.org/10.1016/j.ijheatmasstransfer.2012.09.004

Chapter 9
Design and Optimization of a Thermoelectric Generator for Waste Heat Recovery from Industrial Kiln

Hanin Afifah Ahmad Syukri, Faiza M. Nasir, and Siti Lydia Rahim

Abstract This study focuses on the application of a thermoelectric generator (TEG) tailored explicitly for the purpose of waste heat recovery from industrial kilns. To address this, the study proposes the integration of a TEG, a semiconductor device that effectively converts thermal energy into electrical energy, utilizing the Seebeck effect. Three multi-TEG system configurations were designed and modelled and their performance in recovering the waste heat was simulated using MATLAB Simulink. These configurations encompassed the series arrangement, the parallel arrangement, and the hybrid (series–parallel) arrangement. The outcomes of the simulations conclude the series arrangement as the most effective configuration for the multi-TEG system, showcasing a superior power output of 255 W and an efficiency of 52.26%.

Keywords TEG · MATLAB Simulink · Waste heat recovery · Power · Efficiency

9.1 Introduction

Waste heat recovery offers a valuable alternative for harnessing heat that dissipates within a system during various processes, effectively reutilizing this lost heat through the activation of abandoned energy sources to either produce additional heat or generate electrical and mechanical power. Thermoelectric generators (TEGs) present

---

H. A. A. Syukri
Kulim Hi-Tech Park, Universiti Kuala Lumpur Kampus Cawangan Malaysian Spanish Institute, Kulim, 09000 Kedah, Malaysia
e-mail: hanin.syukri02@s.unikl.edu.my

F. M. Nasir (✉) · S. L. Rahim
Centre for Women Advancement and Leadership, Universiti Kuala Lumpur, 50250 Kuala Lumpur, Wilayah Persekutuan, Malaysia
e-mail: faiza@unikl.edu.my

S. L. Rahim
e-mail: sitilydia@unikl.edu.my

T. A. A. Razak et al. (eds.), *Progress in Engineering Technology VI*, Advanced Structured Materials 229, https://doi.org/10.1007/978-3-031-81517-1_9

a viable solution for recuperating the heat energy lost to the surroundings, capitalizing on thermal energy, which is abundant and prevalent across a multitude of industrial domains. Functioning as semiconductor devices, TEGs possess the inherent ability to directly transform temperature disparities between surfaces into electrical energy through the utilization of the Seebeck effect principle [1].

TEG finds practical application in various domains, including its integration within automobile engines, as highlighted in the research conducted by Mostafavi and Mahmoudi [1]. Additionally, Guan et al. [2] conducted a study emphasizing the utilization of TEG in micro self-powered wireless platforms. Within this research, the microcontroller unit and wireless sensor node were engineered to operate on low input voltage and power, ensuring high efficiency to effectively harvest thermoelectric energy within the system. The versatility of TEGs extends to their application in health monitoring and tracking systems, as affirmed by Thielen et al. [3]. Their study demonstrated the utilization of human body heat to power wearable devices, showcasing the potential of TEG as an independent and viable harvesting source amidst various alternatives such as light, vibration, and radio frequency.

Various configurations of TEG systems have undergone rigorous testing using MATLAB Simulink, as detailed in the work by Ji [4]. In this study, the comparison of series and parallel configurations revealed that the series–parallel configuration yielded the highest levels of power, voltage, and current output. Furthermore, a study [5] delved into a key parameter of this TEG system—temperature difference.

TEG finds versatile applications in waste heat recovery across various domains [6]. However, it is worth noting that the efficiency of TEG-to-electricity conversion has not yet reached desired levels. Hence, it is the main objective of this study to investigate and optimize the design of the TEG for waste heat recovery using MATLAB Simulink.

9.2 Methodology

See Figs. 9.1 and 9.2.

9.2.1 The Single TEG Model

Figure 9.1 illustrates the MATLAB Simulink model representing a straightforward TEG system, comprising a single TEG component. A timer mechanism has been incorporated into this setup to ensure that system parameter values undergo temporal variation. To enhance clarity and simplify the input–output representation of the TEG system, a subsystem block is employed, encapsulating essential Simulink blocks for key parameters, such as Seebeck effect and resistance load, meticulously arranged as depicted in Fig. 9.2. Broadly, this system performs computations involving critical parameters including temperature differential, Seebeck coefficient, resistance,

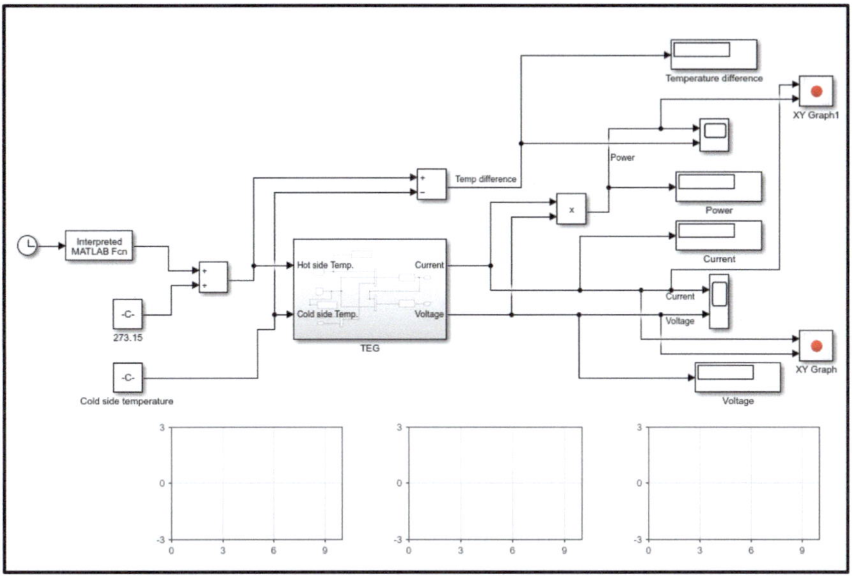

Fig. 9.1 MATLAB Simulink model for simple TEG system

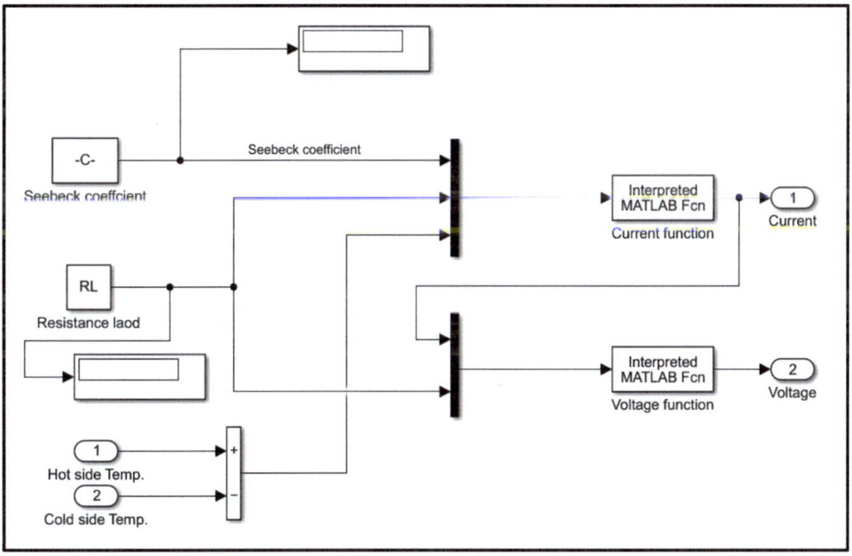

Fig. 9.2 TEG subsystem

Fig. 9.3 TEG in series arrangement

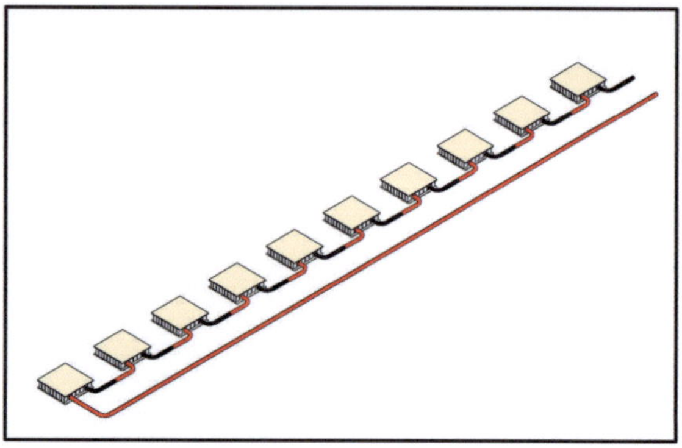

Fig. 9.4 TEG in parallel arrangement

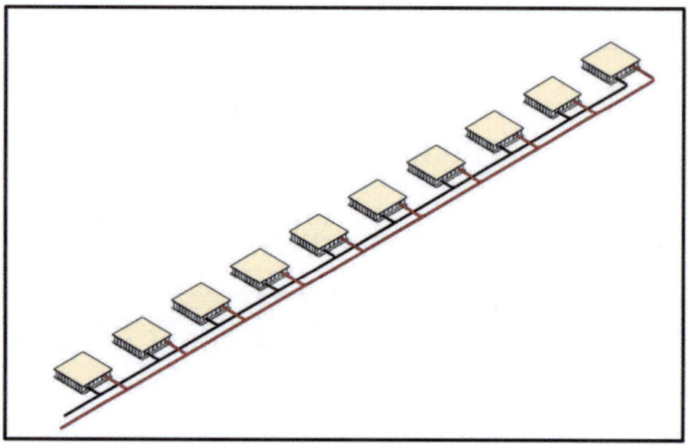

current, voltage, and power of the TEG system, considering hot and cold side temperatures as the primary inputs.

9.2.2 The Multi-TEG System Model

The multi-TEG system was modelled using MATLAB Simulink using the same approach taken for modelling the single TEG system. Three multi-TEG systems with different TEG arrangements were considered: the series arrangement, the parallel arrangement, and the hybrid arrangement. The schematic drawings of the arrangements are shown in Figs. 9.3, 9.4 and 9.5.

Model Building and Simulation for Multi-TEG

In Fig. 9.6, a MATLAB Simulink model for multi-TEG system modelling is presented, building upon the initial model showcased in Fig. 9.1. This improved

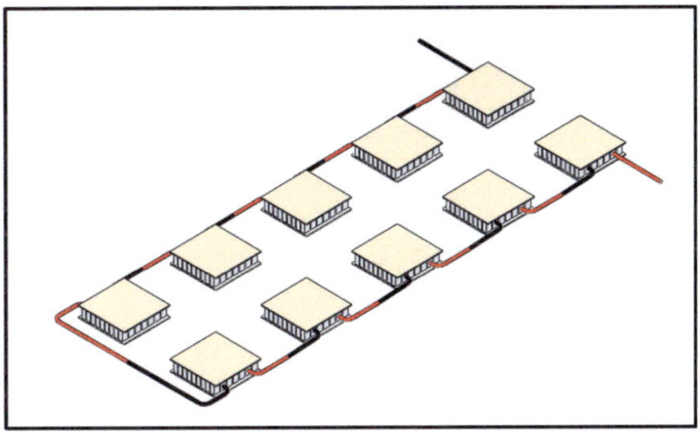

Fig. 9.5 TEG in series–parallel arrangement

model provides comprehensive results, encompassing Seebeck coefficient, ZT value, current, voltage, resistance, power, heat transfer on the hot side (Q_h), and thermal efficiency of the TEG system. The model's input factors include the number of TEGs and the temperatures on the hot and cold sides. The TEG system modelling remains consistent for series, parallel, and series–parallel arrangements, with distinctions occurring within the subsystem due to unique parameter calculations for each configuration.

Figure 9.7 shows TEG subsystem for ten number of TEG in series and Fig. 9.8 shows the subsystem for multi-TEG in parallel configuration. In Fig. 9.9, the current,

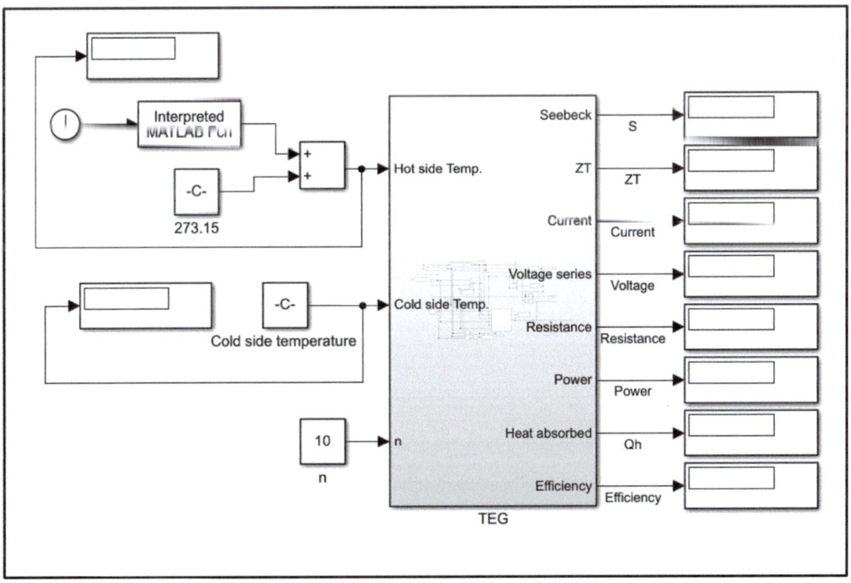

Fig. 9.6 MATLAB Simulink model for multi-TEG system

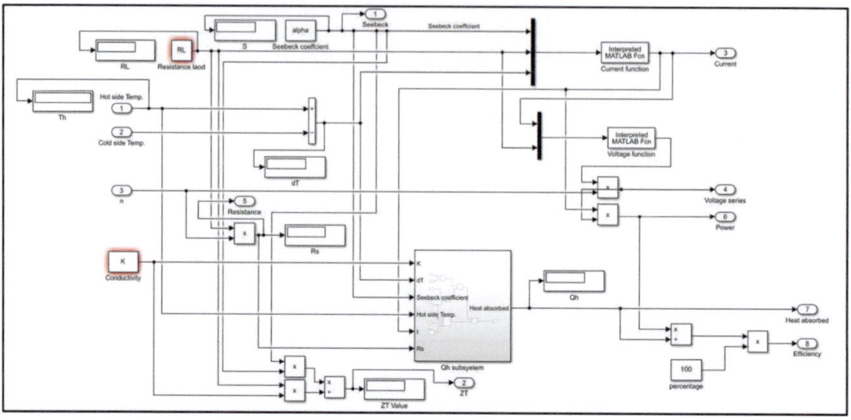

Fig. 9.7 TEG subsystem for series arrangement

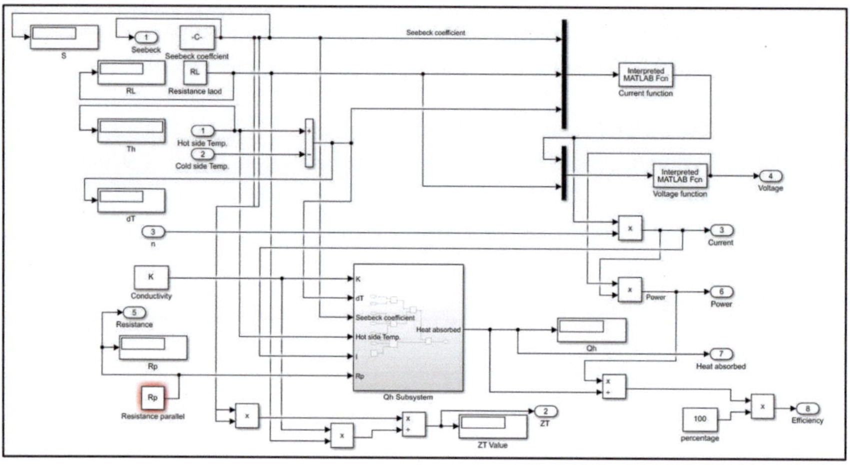

Fig. 9.8 TEG subsystem for parallel arrangement

voltage, resistance are all calculated based on the TEG arrangements in which there are five TEGs in series meaning that there are two rows of TEGs in parallel.

9.3 Result and Discussion

Figure 9.10 shows the current versus time plot for series, parallel, and series–parallel arrangement for the multi-TEG system that was generated from the MATLAB Simulink simulation. Series arrangement maximum current value is 4.554 A, parallel arrangement maximum current value is 45.54 A while series–parallel arrangement

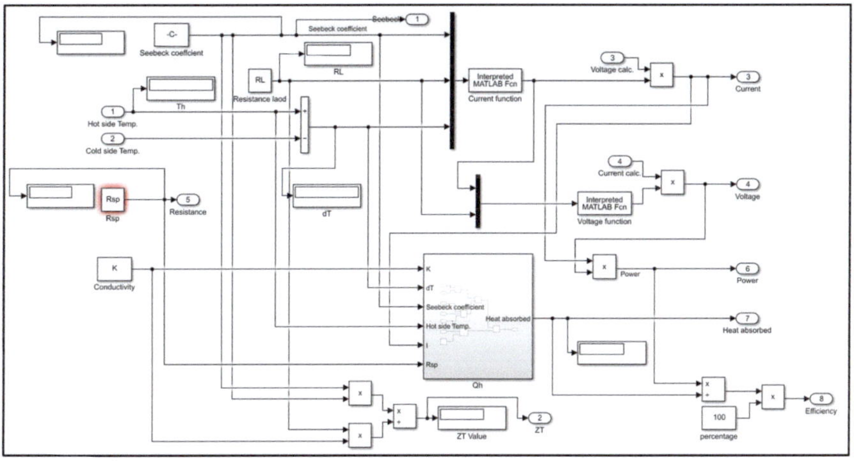

Fig. 9.9 TEG subsystem for series–parallel arrangement

current value is 9.107 A. The current for the TEG system increases linearly for all three arrangements. The highest current value obtained based on the graph is for the parallel arrangement and the lowest is for the series arrangement.

Figure 9.11 depicts the voltage from MATLAB Simulink simulation results for all three arrangements. The minimum value for voltage in series is 56 V, for parallel is 5.6 V, and for series–parallel arrangement is 28 V. The series arrangement recorded the highest voltage value while the parallel arrangement recorded the lowest value.

Figure 9.12 displays the results of graph efficiency of the TEG system from MATLAB Simulink simulation. The series arrangement has the highest efficiency

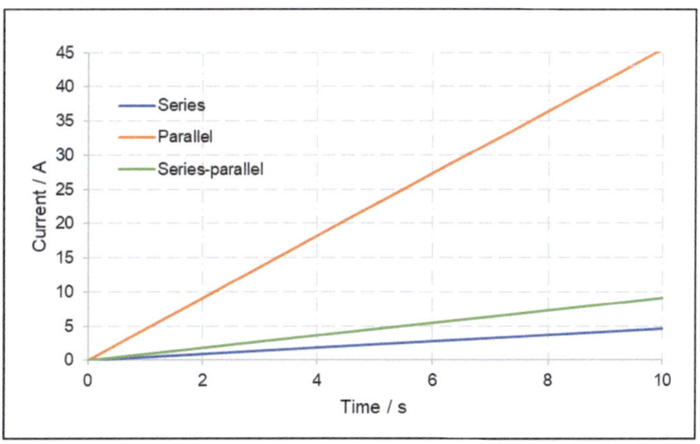

Fig. 9.10 Current versus time for 10 TEG in series, parallel, and series–parallel arrangement

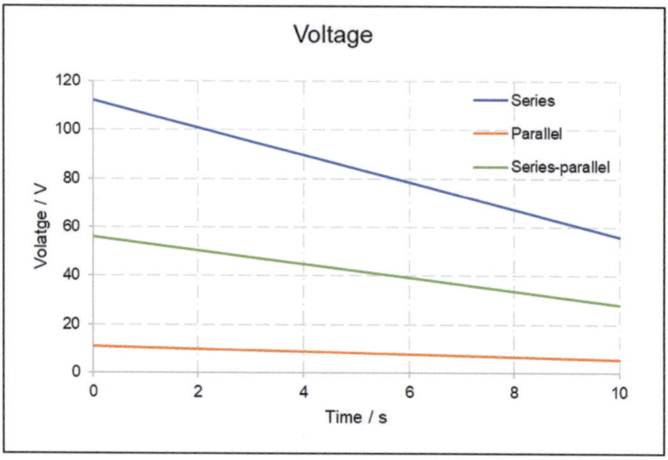

Fig. 9.11 Voltage versus time for 10 TEG in series, parallel, and series–parallel arrangement

while the parallel arrangement TEG system recorded the lowest efficiency of all three designs (Fig. 9.12).

Table 9.1 summarizes all the simulation results for the three multi-TEG systems. After taking the power output and efficiency into account for the final design selection, the series arrangement multi-TEG system is chosen as the optimized final design as it produces the highest thermal efficiency.

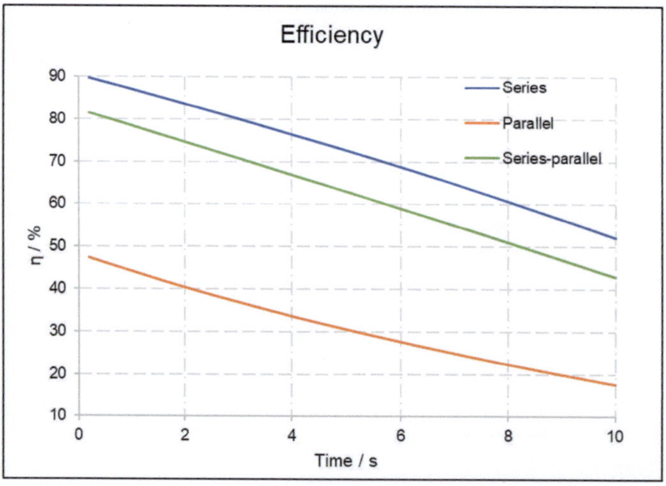

Fig. 9.12 Thermal efficiency versus time for 10 TEG in series, parallel, and series-parallel arrangement

Table 9.1 Summary of MATLAB Simulink simulation data for multi-TEG system

Parameters	Series	Parallel	Series–Parallel
Current, I (A)	4.554	45.54	9.108
Voltage, V (V)	56	5.6	28
Resistant, R (Ω)	12.3	0.123	3.075
Power (W)	255	255	255
Heat transfer, Qh (W)	488	1441	593.9
Efficiency, η (%)	52.26	17.69	42.94

9.4 Conclusions

In conclusion, this study proposes a thermoelectric generator system for waste heat recovery from an industrial kiln. Modelling and simulation of the TEG system were performed using MATLAB Simulink. MATLAB Simulink systems were developed for multi-TEG systems across the three different arrangements, yielding efficiencies of 52.26%, 17.69%, and 42.94% for series, parallel, and series–parallel systems, respectively. With the highest thermal efficiency, the series arrangement with ten TEG units was identified as the optimal final design.

References

1. Hewawasam LS, Jayasena AS, Afnan MMM, Ranasinghe RACP, Wijewardane MA (2020) Waste heat recovery from thermo-electric generators (TEGs). Energy Rep 6:474–479. https://doi.org/10.1016/j.egyr.2019.11.105
2. Guan M, Wang K, Xu D, Liao WH (2017) Design and experimental investigation of a low-voltage thermoelectric energy harvesting system for wireless sensor nodes. Energy Convers Manag. https://doi.org/10.1016/j.enconman.2017.01.049
3. Thielen M, Sigrist L, Magno M, Hierold C, Benini L (2017) Human body heat for powering wearable devices: from thermal energy to application. Energy Convers Manag. https://doi.org/10.1016/j.enconman.2016.11.005
4. Ji D (2018) Modeling and optimization of thermoelectric generator for waste heat recovery
5. Mostafavi SA, Mahmoudi M (2018) Modeling and fabricating a prototype of a thermoelectric generator system of heat energy recovery from hot exhaust gases and evaluating the effects of important system parameters. Appl Therm Eng. https://doi.org/10.1016/j.applthermaleng.2018.01.018
6. Zaferani SH, Jafarian M, Vashaee D, Ghomashchi R (2021) Thermal management systems and waste heat recycling by thermoelectric generators—an overview. Energies (Basel). 14. https://doi.org/10.3390/en14185646

Chapter 10
Design Optimization of an Organic Rankine Cycle with Different Operating Parameters for Heat Recovery from Industrial Kiln

Nisha Hannani Mohd. Hadzari, Faiza Mohamed Nasir, and Siti Lydia Rahim

Abstract This study optimizes an organic Rankine cycle (ORC) system for efficient heat recovery from industrial kilns. Using the DWSIM software for design and simulation, three ORC designs are proposed, with the most promising selected based on power generation and thermal efficiency. The chosen design is analyzed under varying evaporating and condensing pressures to assess their impact. Results show these parameters significantly affect power generation and efficiency. The study concludes that ORC systems effectively recover heat and generate electricity from industrial kilns.

Keywords DWSIM · Waste heat recovery system · Operating parameters · ORC

10.1 Introduction

Waste heat recovery (WHR) technologies are being introduced to achieve the highest possible waste heat recovery efficiency [1]. The organic Rankine cycle (ORC) is a well-studied and widely applied technology for waste heat recovery and power generation. It has been successfully used in various fields, including internal combustion engines [2], gas turbines [3], and nuclear reactors [4]. The ORC operates on the same

N. H. Mohd. Hadzari
Malaysian Spanish Institute Kulim Hi-Tech Park, Universiti Kuala Lumpur Kampus Cawangan, Kulim, 09000 Kedah, Malaysia
e-mail: nisha.hadzari18@s.unikl.edu.my

F. M. Nasir (✉) · S. L. Rahim
Centre for Women Advancement and Leadership, Universiti Kuala Lumpur, 50250 Kuala Lumpur, Wilayah Persekutuan, Malaysia
e-mail: faiza@unikl.edu.my

S. L. Rahim
e-mail: sitilydia@unikl.edu.my

T. A. A. Razak et al. (eds.), *Progress in Engineering Technology VI*, Advanced Structured Materials 229, https://doi.org/10.1007/978-3-031-81517-1_10

principle as the traditional Rankine cycle, but it utilizes organic working fluids with lower boiling points, enabling operation at lower temperatures [5].

There are six types of ORC which are the basic ORC (BORC), single-stage regenerative ORC (SRORC), double-stage regenerative ORC (DRORC), reheat ORC (RORC), ORC with recuperator and dual-loop ORC (DLORC). The performance of the ORC is influenced by various operating parameters. Wei et al. [3] conducted sensitivity analyses of key parameters and found that integrating the ORC with an intercooled cycle gas turbine can increase the output power and efficiency of the turbine. The selection of organic working fluids also significantly affects the performance of the ORC. Touaibi et al. [5] compared the performance of the ORC using three different organic fluids (toluene, R245fa, and R123) and concluded that the choice of working fluid has a significant impact on the cycle performance. Heat exchangers play a crucial role in the performance of the ORC. Erdogan and Colpan [6] assessed the effect of source temperature on the performance of the heat exchanger and the ORC using a system-level model.

Overall, the ORC is a versatile technology with various applications and configurations. Its performance is influenced by operating parameters, working fluid selection, and heat exchanger design. Further research is needed to optimize the performance of the ORC and explore its potential in different fields.

10.2 Methodology

This study uses a methodology that includes design, simulation, basic thermoeconomic analysis, and parametric studies of the operating conditions of ORC. The DWSIM software was used to simulate the ORC system and to produce the power generated and the thermal efficiency of the ORC. Upon validation of the ORC model in the DWSIM, three conceptual designs of ORC were designed, and the best design was selected based on the power generated and thermal efficiency. Then parametric studies of the operating parameters on the ORC performance were conducted.

10.2.1 The Models

A benchmark flowsheet model of an ORC plant was developed using the DWSIM (Fig. 10.1) based on the BORC model. The model was based on the Chena geothermal power plant described in the study of Aneke et al. [7].

The validation of the DWSIM model was performed by comparing the simulation results with the real plant design data, the study was able to validate the accuracy of the simulation model and ensure that the results were reliable. The simulated results were compared against the actual plant data presented in the study of Aneke et al. [7]. The comparison is shown in Table 10.1. The simulation results differ less than

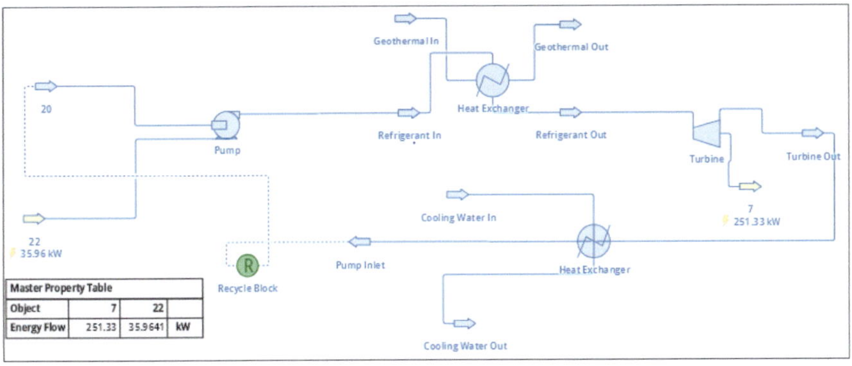

Fig. 10.1 DWSIM model of the Chena geothermal power plant

Table 10.1 Comparison between real power plant data and DWSIM simulation results

Parameter	Real power plant data	DWSIM simulation result	% difference
Cooling water exit temperature (° C)	10	9.91	0.90%
Net plant power (kW)	210	230.83	9.92%
Thermal efficiency	8.14	8.44	3.69%
Evaporator heat transfer rate (kW)	2580	2551.62	1.10%
Condenser heat transfer rate (kW)	2360	2336.26	1.01%

10% from the real power plant data and hence the DWSIM model is considered to be a valid model.

Two other DWSIM flowsheet models were developed using the same validated approach for the BORC model. The models were based on the configuration of SRORC and DEORC. Their thermal efficiencies were then compared with that of the benchmark model (the BORC).

10.2.2 Thermoeconomic Analysis

Thermoeconomic analysis is used to evaluate the economic and thermodynamic performance of energy systems by analyzing the cost of producing useful work and the cost of resources consumed in the process [8]. The costs that were involved in thermoeconomic calculation were the total equipment cost, total module cost, annual utility cost, total power generated annual (kWh) and profit. Each component of bare module cost (CBM) was determined using the module costing approach, and the purchased cost was determined using the chemical engineering plant cost index (CEPCI) [9]. Thermoeconomic analyses were also used to determine the PBP of the three ORC configuration designs developed in the DWSIM (BORC, SRORC and DEORC). The ORC layout with the best thermal efficiency and the lowest PBP is then used for the subsequent parametric studies.

10.2.3 *Parametric Studies*

The final ORC design and layout derived in Sect. 2.2 undergo further optimization concerning various operating parameters. These parameters encompass condensing pressure and evaporating pressure. In this specific case study, certain plant parameters were held constant while others were allowed to vary, mirroring a real plant system.

10.3 Results and Discussion

See Fig. 10.2 and Table 10.2.

10.3.1 *Comparison Between the ORC Configurations*

Table 10.2 shows the performance comparison between the three ORC designs/configurations. The DEORC (design 3) exhibits superior thermal efficiency compared to BORC (design 1) and SRORC (design 2), with thermal efficiencies of 19.71%, 8.44%, and 18.7%, respectively. Design 2 emerges as the optimal choice due to its superior efficiency percentage compared to both design 1 and design 3. Despite the inclusion of two pumps in design 2, it achieves the highest efficiency.

Table 10.2 Comparison of thermal performance for the three ORC designs

Parameters	BORC (benchmark design 1)	SRORC (design 2)	DEORC (design 3)
Evaporator heat transfer (kW)	2551.62	1190	1404.43
Turbine power generated (kW)	251.33	278.57	331.76
Pump power (kW)	35.96	56.2	54.94
Net power plant generated by ORC (kW)	230.83	222.37	276.82
Thermal efficiency (%)	8.44	18.7	19.71

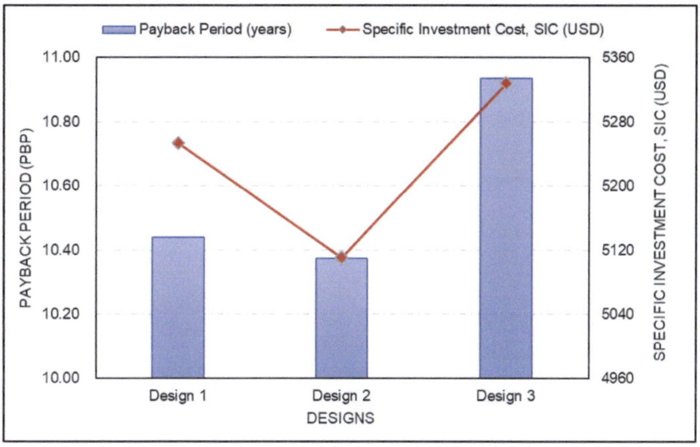

Fig. 10.2 The payback period and specific investment cost for the three ORC designs

From Fig. 10.2, design 2 exhibits the shortest payback period at 10.37 years, closely followed by design 1 at 10.44 years and design 3 at 10.94 years, showcasing minimal disparity between the designs. Among the designs, design 2 showcases the most favorable SIC at USD5,110.61. It is evident that design 2 (SRORC) stands out with the lowest PBP and SIC.

10.3.2 Parametric Studies

Design 2 of the ORC has been proven to be the optimal design between the three ORC configurations being studied in this work. The design undergoes parametric studies to establish the effect of operating parameters on the performance of the ORC, i.e. the thermal efficiency, the total power generated by the turbine, the payback period and the specific investment cost.

Evaporating Pressure

In Fig. 10.3, it can be observed that the higher the evaporating pressure, the lower is the thermal efficiency and the power generated by the turbine. Specifically, at 12.75 bar, the highest efficiency and total turbine output are observed, reaching 27% and 519.93 kW, respectively. The findings in Fig. 10.4 illustrate a linear proportionality between payback the period (PBP) and specific investment cost (SIC). The PBP and the SIC increase linearly with the evaporating pressure. Specifically, at 19.13 bar, the PBP and SIC reach their highest values, while at 12.75 bar, the PBP and SIC are at their lowest.

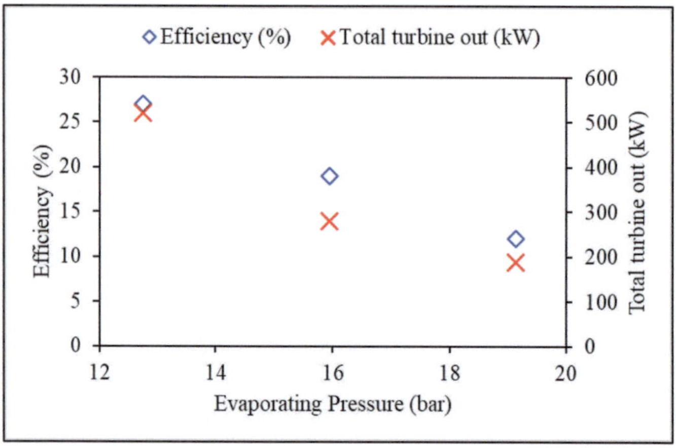

Fig. 10.3 The effect of evaporating pressure on the thermal performance of the SRORC

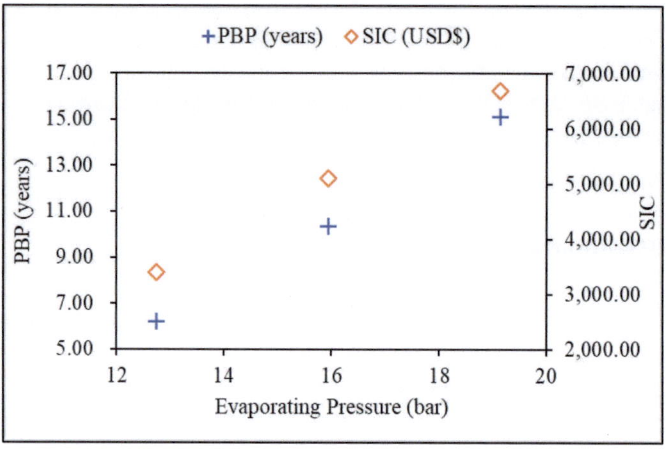

Fig. 10.4 The effect of evaporating pressure on the PBP and SIC of the SRORC

Condensing Pressure

In Fig. 10.5, the thermal efficiency (%) and the total generated power are plotted against varying condensing pressures. The higher the condensing pressure, the lower is the thermal performance of the SRORC. The highest performance of SRORC can be attained when the condensing pressure is at 1.76 bar, yielding thermal efficiency of 27% and generated power of 447.16 kW.

Figure 10.6 illustrates the payback period and the specific investment cost for different condensing pressures. A favorable economic performance of the ORC can

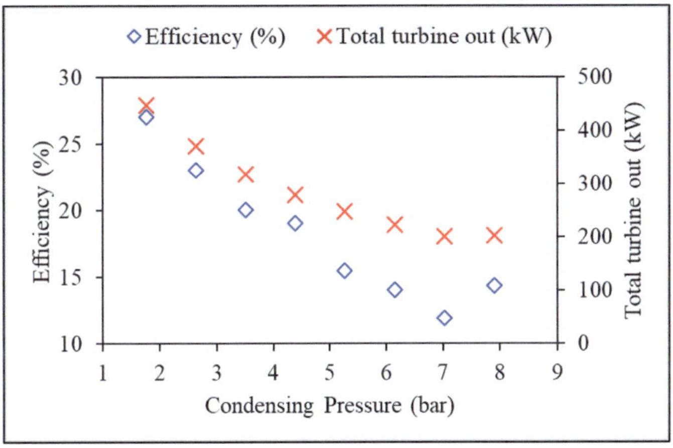

Fig. 10.5 The effect of condensing pressure on the thermal performance of the SRORC

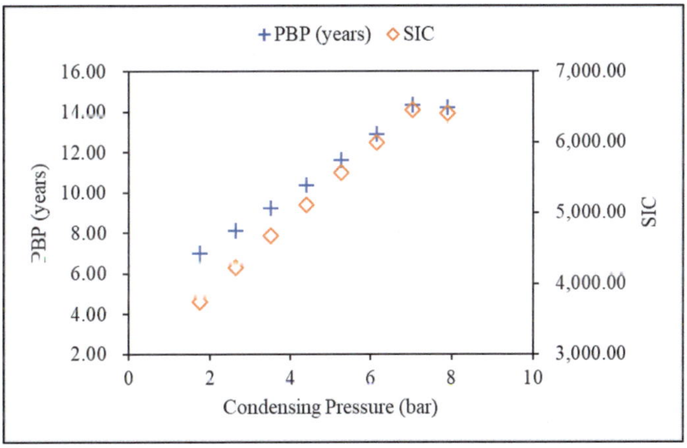

Fig. 10.6 The effect of condensing pressure on the PBP and SIC of the SRORC

be obtained when the condensing pressure is at its possible lowest value. Particularly, at the lowest pressure of 1.76 bar, the PBP is 6.99 years with an SIC of USD$3745.02.

10.4 Conclusions

In this study, an ORC system was successfully modeled and validated using data from the Chena geothermal power plant, affirming the software's capability to simulate ORCs with commendable accuracy. Various designs, such as BORC, SRORC, and DEORC, were explored, each yielding distinct outcomes in power generation and thermal efficiency. Through subsequent thermoeconomic analysis, it was concluded that design 2 (SRORC) emerged as the optimal choice for this study. The thermal performance and thermoeconomic analysis of the ORC system were evaluated based on the power generated, thermal efficiency, payback period, and total initial investment cost (Fig. 10.6).

Parametric studies were conducted by varying the evaporating pressure and condensing pressure of the working fluid for the SRORC while keeping other plant parameters fixed. It can be concluded that the evaporating pressure is the most important parameter affecting the SRORC's efficiency and payback period.

References

1. Jouhara H, Khordehgah N, Almahmoud S, Delpech B, Chauhan A, Tassou SA (2018) Waste heat recovery technologies and applications
2. Tahani M, Javan S, Biglari M (2013) A comprehensive study on waste heat recovery from internal combustion engines using organic rankine cycle. Therm Sci. https://doi.org/10.2298/TSCI111219051T
3. Liu W, Zhang X, Zhao N, Shu C, Zhang S, Ma Z, Han J (2018) Performance analysis of organic Rankine cycle power generation system for intercooled cycle gas turbine. Adv Mech Eng. https://doi.org/10.1177/1687814018794074
4. Shokati N, Mohammadkhani F, Yari M, Mahmoudi SMS, Rosen MA (2014) A comparative exergoeconomic analysis of waste heat recovery from a gas turbine-modular helium reactor via organic rankine cycles. Sustainability (Switzerland). https://doi.org/10.3390/su6052474
5. Touaibi R, Koten H, Boydak O (2020) Parametric study of an organic rankine cycle using different fluids. Emerg Sci J. https://doi.org/10.28991/esj-2020-01216
6. Erdogan A, Colpan CO (2018) Performance assessment of shell and tube heat exchanger based subcritical and supercritical organic Rankine cycles. Therm Sci 22. https://doi.org/10.2298/tsci171101019e
7. Aneke M, Agnew B, Underwood C (2011) Performance analysis of the Chena binary geothermal power plant. Appl Therm Eng 31:1825–1832. https://doi.org/10.1016/j.applthermaleng.2011.02.028
8. Valencia G, Duarte J, Isaza-Roldan C (2019) Thermoeconomic analysis of different exhaust waste-heat recovery systems for natural gas engine based on ORC. Appl Sci (Switzerland) 9. https://doi.org/10.3390/app9194017
9. Song J, Loo P, Teo J, Markides CN (2020) Thermo-economic optimization of organic rankine cycle (ORC) systems for geothermal power generation: a comparative study of system configurations

Chapter 11
The Effect of Different Working Fluids on the Thermal and Economic Performance of Organic Rankine Cycles for Heat Recovery from Industrial Kiln

Nursyuhadah Othman and Faiza Mohamed Nasir

Abstract Waste heat recovery (WHR) involves capturing and utilizing heat energy that would otherwise be dissipated into the environment. Numerous studies indicated that the organic Rankine cycle (ORC) is a promising approach for waste heat recovery. The primary goal of this study is to refine the operational aspects of the ORC system to efficiently capture and convert heat from industrial kilns into electricity. The methodology involves a comprehensive approach encompassing design, simulation, and fundamental thermoeconomic analysis for the ORC system. The DWSIM software is utilized to simulate the ORC system and assess the resulting outcomes. The study analyzes three distinct conceptual designs of the ORC system, ultimately selecting the most promising design based on power generation, thermal efficiency, and payback period. This study also explores the influence of diverse working fluids on the thermal and economic performance of the ORC. Through thermoeconomic analysis and consideration of payback periods, the optimal design and the most suitable working fluid, R134a (tetrafluoroethane), were identified.

Keywords Waste heat energy · Industrial calcination · ORC · Thermal efficiency · Working fluids

---

N. Othman
Dialog E&C Sdn Bhd, Dialog Tower, No 15 Jalan PJU 7/5, Mutiara Damansara, 47810 Petaling Jaya, Selangor Darul Ehsan, Malaysia
e-mail: nursyuhadah.othman@dialogasia.com

F. M. Nasir (✉)
Universiti Kuala Lumpur Kampus Cawangan Malaysian Spanish Institute Kulim Hi-Tech Park, 09000 Kulim, Kedah, Malaysia
e-mail: faiza@unikl.edu.my

T. A. A. Razak et al. (eds.), *Progress in Engineering Technology VI*, Advanced Structured Materials 229, https://doi.org/10.1007/978-3-031-81517-1_11

11.1 Introduction

The increasing demand for energy raises concerns about potential strain on limited natural resources essential for sustaining industrial societies. Waste heat recovery (WHR) addresses this concern by capturing thermal energy that would otherwise dissipate into the environment and utilizing it efficiently. The organic Rankine cycle (ORC) stands as an excellent solution for generating significant energy at relatively lower temperatures, reducing the dependency on fossil fuels. It has found successful applications across various domains, including internal combustion engines [1], gas turbines [2], and nuclear reactors [3]. Operating on the same principle as the traditional Rankine cycle, the ORC utilizes organic working fluids with lower boiling points, enabling operations at lower temperatures [4].

The configuration of the ORC system plays a pivotal role in optimizing its performance. Six primary ORC configurations exist, including the basic ORC (BORC), single-stage regenerative ORC (SRORC), double-stage regenerative ORC (DRORC), reheat ORC (RORC), ORC with recuperator, and dual-loop ORC (DLORC). The effectiveness and functionality of an ORC system are fundamentally influenced by the choice of the working fluid and its adherence to defined thermodynamic, environmental, and safety criteria [5]. Notably, Douvartzides and Karmalis [6] undertook an extensive evaluation involving 37 distinct working fluid substances. Their findings highlighted that judicious selection of the working fluid, along with appropriately tailored operational parameters for the cycle, could potentially enhance the overall plant efficiency by nearly 6% and concurrently reduce fuel consumption by up to 13%. Touaibi et al. [4] compared the performance of the ORC using three different organic fluids (toluene, R245fa, and R123) and concluded that the choice of working fluid substantially influences cycle performance.

The main objective of this study is to investigate the effect of different working fluids on the thermal and economic performance of the ORC in recovering waste heat from industrial applications.

11.2 Methodology

This study employs a comprehensive methodology involving design, simulation, fundamental thermoeconomic analysis and systematic investigation of operational conditions for an organic Rankine cycle (ORC) system. The simulation of the ORC system and the subsequent assessment of power generation and thermal efficiency were conducted using DWSIM software. Following the validation of the ORC model within the DWSIM environment, three distinct conceptual ORC designs were formulated. The selection of the optimal design was based on criteria such as power generation and thermal efficiency. Subsequently, parametric studies exploring the influence of working fluids on the ORC performance were conducted.

11.2.1 The Model

A benchmark flowsheet model representing an organic Rankine cycle plant was constructed using DWSIM, as depicted in Fig. 11.1, building upon the basic organic Rankine cycle. The benchmark design for this model was adapted from the Chena geothermal power plant, detailed in the work of Aneke et al. [7].

Validation of the DWSIM model was accomplished by meticulously comparing the simulation outcomes with the authentic design data of the real plant. The simulated results were meticulously compared to the plant data, detailed in the work of Aneke et al. [7], and the comparison findings have been tabulated in Table 11.1. It is noteworthy that the deviation between the simulation results and actual plant data was within a margin of less than 10%. Consequently, the DWSIM model was deemed valid and reflective of the real power plant.

Two additional DWSIM flowsheet models were generated utilizing the validated methodology established for the basic organic Rankine cycle model. These models

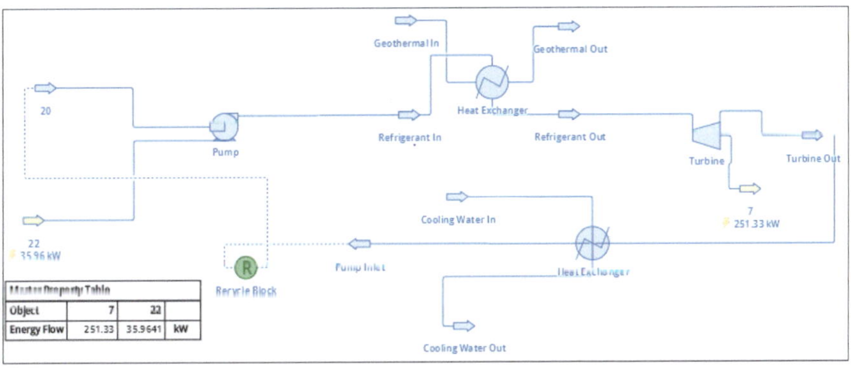

Fig. 11.1 DWSIM model of the Chena geothermal power plant

Table 11.1 Comparison between real power plant data and DWSIM simulation results

Parameter	Real power plant data	DWSIM simulation result	Percentage difference (%)
Cooling water exit temperature (°C)	10	9.91	0.90
Net plant power (kW)	210	230.83	9.92
Thermal efficiency (%)	8.14	8.44	3.69
Evaporator heat transfer rate (kW)	2580	2551.62	1.10
Condenser heat transfer rate (kW)	2360	2336.26	1.01

were structured according to the single-stage regenerative organic Rankine cycle configuration and the double-expansion organic Rankine cycle configuration.

11.2.2 Thermoeconomic Analysis

The thermoeconomic analysis encompassed various cost parameters, including total equipment cost, total module cost, annual utility cost, total annual power generation (in kWh), and profit. To calculate each component's bare module cost (CBM), the module costing approach was employed. Additionally, the purchased cost was determined using the chemical engineering plant cost index (CEPCI) [8].

Payback period is the time expected to return an investment's expenses or for an investor to break even. Longer payback periods are undesirable, whereas shorter payback periods are more tempting investments. The payback period (PBP), which can be determined as below, is chosen as the indicator to assess ORC systems from an economic perspective and also for economic optimization [9].

In this study, thermoeconomic analysis was employed to calculate the PBP for the three distinct ORC configuration designs developed within DWSIM, namely BORC, SRORC, and DEORC. Subsequently, the ORC layout demonstrating the optimal thermal efficiency and the shortest PBP was identified and this configuration was utilized to determine the effect of using different working fluids in the ORC.

11.2.3 Effect of Different Working Fluids

It is the main objective of this study to determine the effect of ORC working fluids on the thermal and economic performance of the ORC. This comparative study employs the selected ORC configuration from the previous section, involving five types of working fluid. They are R134a, R123, R218, R245fa, and R113.

11.3 Result and Discussion

11.3.1 Comparison Between the ORC Designs

Table 11.2 presents a comparative analysis of performance across the three ORC designs/configurations. Design 3, or DEORC, demonstrates notably higher thermal efficiency (19.71%) when juxtaposed with design 1 (BORC) at 8.44% and design 2 (SRORC) at 18.7%. Furthermore, design 1 exhibits the least turbine power output at 251.33 kW, followed by design 2 at 278.57 kW, while design 3 displays the highest turbine power output at 331.76 kW. Design 2 emerged as the optimal choice due to

Table 11.2 Comparison of thermal performance for the three ORC designs

Parameters	BORC (benchmark design 1)	SRORC (design 2)	DEORC (design 3)
Evaporator heat transfer (kW)	2551.62	1190	1404.43
Turbine power generated (kW)	251.33	278.57	331.76
Pump power (kW)	35.96	56.2	54.94
Net power plant generated by ORC (kW)	230.83	222.37	276.82
Thermal efficiency (%)	8.44	18.7	19.71

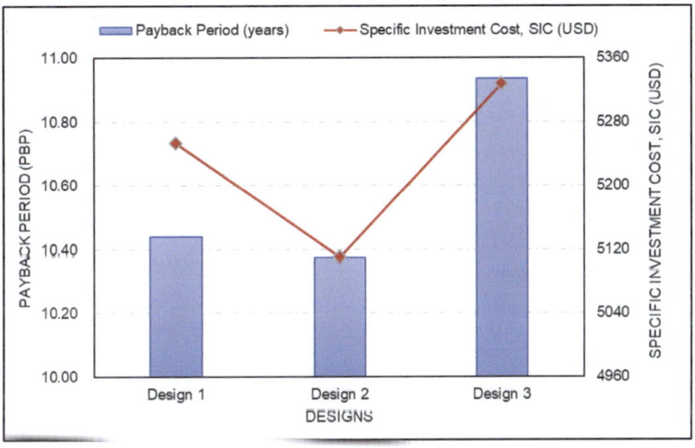

Fig. 11.2 The payback period and specific investment cost for the three ORC designs

its higher efficiency percentage compared to both design 1 and design 3. Despite the inclusion of two pumps in design 2, it achieved the highest efficiency.

Figure 11.2 illustrates the outcomes of thermoeconomic analysis for the three ORC designs. Design 2 exhibited the shortest payback period at 10.37 years, closely followed by design 1 at 10.44 years and design 3 at 10.94 years, showcasing minimal disparity between the designs. The corresponding SIC for each design is presented in the same Fig. 11.2. It is evident that design 2 (SRORC) stood out with the lowest PBP and SIC.

11.3.2 Effect of Different Working Fluids

The thermal efficiencies for different ORC working fluids are shown in Fig. 11.3. When assessing the range of working fluids under consideration, it was observed that

R123 stood out with the highest thermal efficiency, reaching an impressive score of 24.15%. R245fa emerged as the working fluid with the least thermal efficiency in this analysis, presenting a markedly lower rate of 14.42%.

The thermoeconomic performance of the ORC with different working fluids is highlighted in Figs. 11.4 and 11.5. Figure 11.4 provides a visual representation of the payback periods associated with employing ORC with different working fluids. Notably, R113 exhibits the lengthiest payback period, requiring 32 years to recuperate the initial investment. In sharp contrast, R134a boasts the shortest payback period among the five working fluids, necessitating only 10.3 years to recover the investment.

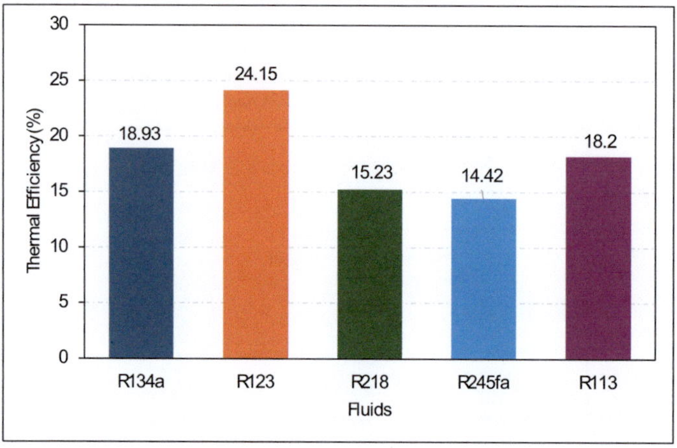

Fig. 11.3 Thermal efficiency of the ORC for different working fluids

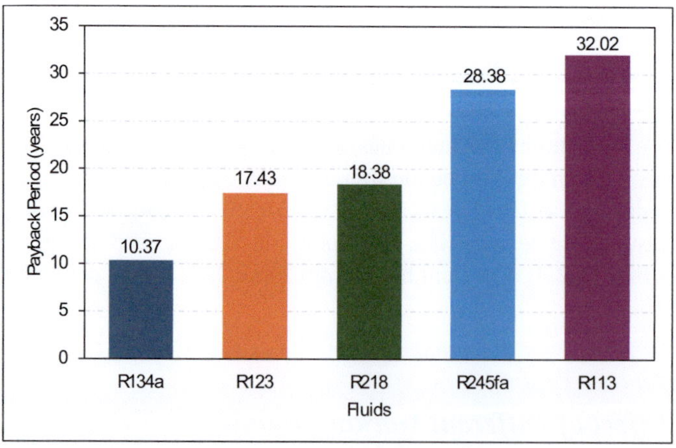

Fig. 11.4 The PBP for the ORC with different working fluids

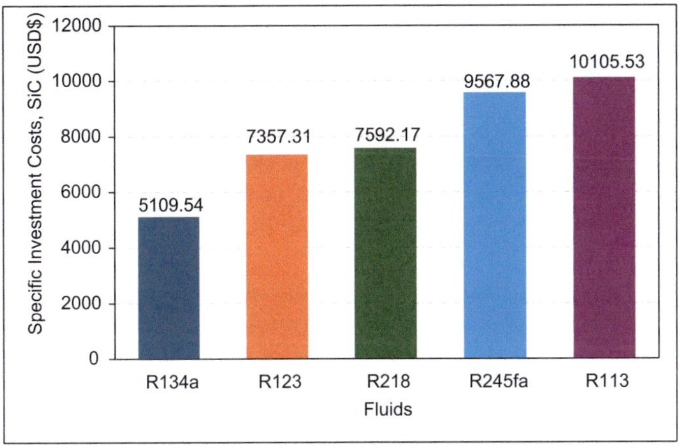

Fig. 11.5 The specific investment cost (SiC) for the ORC with different working fluids

Figure 11.5 delineates the specific investment costs (SiC) attributed to five distinct working fluids. Analyzing the data, it is evident that R113 incurs the highest specific investment cost, amounting to USD$10,105. Remarkably, R134a stands out with the most economically favorable specific investment cost among the working fluids, totaling USD$5,109.54.

A summary of the findings in this study is shown in Table 11.3, where the top performers (i.e. the working fluid) for each assessment criteria are identified. Considering the thermal and economic performance of the working fluid, it can be concluded that the most optimal fluid for ORC is R134a as it gives good thermal efficiency, the highest generated power with the lowest PBP, and SIC.

Table 11.3 Summary of ORC performance

Assessment criteria	Top performer	Second-top performer
Thermal efficiency (%)	R123 (24.15%)	R134a (18.93%)
Total annual generated power (millions kWh)	R134a (2.24)	R123 (1.28)
Payback period (years)	R134a (10.37)	R123 (17.43)
Specific investment cost (USD$ × 1000)	R134a (5.11)	R123 (7.36)

11.4 Conclusions

This study aimed to investigate the performance of ORC systems with various working fluids. Five types of working fluids have been investigated and thermoeconomic analysis also had been conducted for each of the working fluids. The results of the study show that the R134a-based ORC design had the highest thermal efficiency and the lowest PBP and SIC, indicating that it is the most optimal working fluid to be used for ORC.

The study also demonstrates the effectiveness of DWSIM as a simulation tool for ORC design and optimization. The simulation results varied within 10% of the actual real data, indicating that DWSIM is a reliable tool for predicting the performance of ORC systems.

References

1. Tahani M, Javan S, Biglari M (2013) A comprehensive study on waste heat recovery from internal combustion engines using organic Rankine cycle. Therm Sci 17:611–624. https://doi.org/10.2298/TSCI111219051T
2. Liu W, Zhang X, Zhao N, Shu C, Zhang S, Ma Z, Han J (2018) Performance analysis of organic Rankine cycle power generation system for intercooled cycle gas turbine. Adv Mech Eng 10. https://doi.org/10.1177/1687814018794074
3. Shokati N, Mohammadkhani F, Yari M, Mahmoudi SMS, Rosen MA (2014) A comparative exergoeconomic analysis of waste heat recovery from a gas turbine-modular helium reactor via organic Rankine cycles. Sustainability (Switzerland) 6:2474–2489. https://doi.org/10.3390/su6052474
4. Touaibi R, Koten H, Boydak O (2020) Parametric study of an organic Rankine cycle using different fluids. Emerg Sci J 4:122–128. https://doi.org/10.28991/esj-2020-01216
5. Saleh B, Koglbauer G, Wendland M, Fischer J (2007) Working fluids for low-temperature organic Rankine cycles. Energy 32:1210–1221. https://doi.org/10.1016/j.energy.2006.07.001
6. Douvartzides S, Karmalis I (2016) Working fluid selection for the organic Rankine cycle (ORC) exhaust heat recovery of an internal combustion engine power plant. IOP Conf Ser Mater Sci Eng
7. Aneke M, Agnew B, Underwood C (2011) Performance analysis of the Chena binary geothermal power plant. Appl Therm Eng 31:1825–1832. https://doi.org/10.1016/j.applthermaleng.2011.02.028
8. Song J, Loo P, Teo J, Markides CN (2020) Thermo-economic optimization of organic Rankine cycle (ORC) systems for geothermal power generation: a comparative study of system configurations. Front Energy Res 8:1–14. https://doi.org/10.3389/fenrg.2020.00006
9. Shu G, Yu G, Tian H, Wei H, Liang X (2014) A multi-approach evaluation system (MA-ES) of organic Rankine cycles (ORC) used in waste heat utilization. Appl Energy 132:325–338. https://doi.org/10.1016/j.apenergy.2014.07.007

Chapter 12
Design and Development of a Universal Shape Adapting Jig for CNC Milling

Johan Ihsan Mahmood, Adam bin Zulkifli, and Mior Firdaus M. A. Majid

Abstract This paper explains the basics of what a universal shape adapting jig is, how it works and its application. The clamping of the workpiece with the various shapes can be done using this universal shape adapting jig. Testing is done using the clamping test with an appropriate clamping force using air pressure. In CNC milling industries, a lot of jigs had to be produced for parts that could not be properly clamped by conventional milling jigs. Making a universal jig for CNC milling that can adapt to any form is a challenge. This paper evaluates the design of the jig that can adapt to various shapes for CNC milling by using pneumatic cylinders as their clamping tools. The steps involved in implementing this project are conceptual design, detailed CAD design, fabrication and testing. FEA analysis is also conducted to find out the location of the highest stress and deformation that occurs in the design.

Keywords Universal shape adapting jig · CNC milling · Clamping force

12.1 Introduction

Workpieces used for CNC milling process will have a variety of shapes and sizes, hence, this will require various types of jigs to hold the workpiece in place. Irregular side surfaces will hinder the proper clamping of workpieces and require a lot of fine-tuning during the CNC machining process. This is the main motivation to design a universal shape adapting jig that can conform to a variety of workpiece shapes. Jigs and fixtures are very important during machining and manufacturing environment

J. I. Mahmood (✉) · A. Zulkifli · M. F. M. A. Majid
Universiti Kuala Lumpur, Kampus Cawangan Malaysian Spanish Institute Kulim Hi-Tech Park, 09000 Kulim, Kedah, Malaysia
e-mail: johanihsan@unikl.edu.my

A. Zulkifli
e-mail: adam.zulkifli07@s.unikl.edu.my

M. F. M. A. Majid
e-mail: miorfirdaus@unikl.edu.my

T. A. A. Razak et al. (eds.), *Progress in Engineering Technology VI*, Advanced Structured Materials 229, https://doi.org/10.1007/978-3-031-81517-1_12

since it will eliminate the need for a special setup for every workpiece, therefore, facilitating production and ensuring that every workpiece is manufactured within a predetermined tolerance [1]. There are several types of jigs which are typically used in CNC milling; for example, template jigs, which machinists use for precision since the jig is simple and appropriate for their accuracy. Plate jigs work in the same way as template jigs, but unlike template jigs, they have built-in clamps to hold the workpiece. Angle-plate jigs are the most appropriate jigs for drilling at angles [2]. In addition, the manufacturing will reduce the cost and time required to design a specialized jig before starting a machining process.

The objectives of this project are:

- To design and fabricate a universal shape adapting jig that can clamp any shape for CNC milling process.
- To conduct analysis on the universal adapting jig to determine the structural integrity of the jig under load.
- To test the prototype functionality during usage to make sure the workpiece clamp properly with appropriate force.

12.2 Literature Review

12.2.1 Milling Jigs

Typically, workpieces are held steadily during milling with the aid of a mechanical device called a machine vise. Machining vises are used in many common production facilities for tasks demanding high accuracy and close tolerance. Machining vises are frequently more effective in holding a workpiece still during high-speed machining operations. However, the main drawback of the machine vise is that they cannot hold parts with complicated shapes. Hence, specialized custom jigs need to be fabricated to secure the clamping process. There is also no facility for locating and holding free-form objects [3]. Before designing a jig, there are elements that need to be considered thoroughly such as the main body, clamping devices, locating devices and tool guides.

12.2.2 Pneumatic Actuator

Pneumatics focuses on using compressed air to do tasks. Compressed air has undergone procedures to lower its volume and raise its pressure. The typical pressure ranges for using it as a working medium vary from 6 to 8 kg/sq mm. Up to 50 kN of maximum force can be generated by pneumatic systems. The controls can be used manually, electrically or pneumatically. To move objects, a piston or vane is typically powered by compressed air. A pneumatic actuator, often known as an air

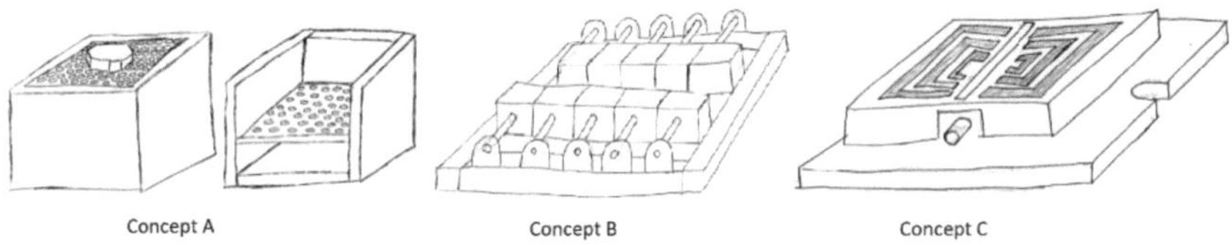

Fig. 12.1 Conceptual designs of the universal shape adapting jig

cylinder, is one of the simplest and most affordable ways to achieve this. An actuator converts a static power source into useful output motion. It can also be used to exert pressure. Typically, actuators are mechanical parts that take energy and transform it into motion. To do this, any movement may be used, such as blocking, clamping or ejection.

12.3 Methodology

12.3.1 Conceptual Design

The development of the jig starts with three conceptual designs A, B and C. Design A is inspired from the universal socket wrench mechanism which has multiple steel cylinders that conform to the shape. Design B consists of multiple pneumatic cylinder actuators that function as clamp tools. Whereas design C will use the vacuum pump to suck the workpiece as the clamping process (Fig. 12.1).

12.3.2 Evaluation Criteria

In order to determine which concept is the best, an evaluation matrix is used on the basis of a few criteria such as ability to clamp any shape, potential to reduce part deformation, ease of manufacturing, manufacturing cost, ability to reduce setup time, ergonomics and maintenance. Ratings are taken from 1 (worst) to 5 (best). Weight factor is also factored in the evaluation criteria, refer to Table 12.1. Design A did not meet a criterion by getting the rating 3 for the first criterion because the mechanism can adapt to a universal shape of part but cannot clamp the part firmly. Apart from that, the potential to reduce part deformation can also be affected because this design cannot clamp firmly. It is also difficult to manufacture, a lot of parts need to be assembled and will increase the manufacturing cost. For design B, it had a good design jig, which can clamp any shape by having several pneumatic cylinders that can easily adapt to the desired shape of the part. Lastly, design C works by using the vacuum pump suction to adhere the workpiece to the machining table. However, the

Table 12.1 Rating matrix table

Criteria	Weight factor	Rating			Weight factor X rating		
		A	B	C	A	B	C
Ability to clamp any shape	5	3	5	4	15	25	20
Reduce workpiece deformation	5	3	5	4	15	25	20
Ease for manufacture	4	2	4	4	8	16	16
Low manufacturing cost	4	5	4	3	20	16	12
Ability to reduce setup time	4	3	4	4	12	16	16
Ergonomics	3	3	3	3	9	9	9
Maintenance	3	2	5	4	6	15	12
		Total			85	122	105

limitation is when the workpiece does not have a flat surface, hence, the clamping force will be poor. Based on these criteria, design B is chosen to be developed since it scored well to the value of 122.

12.3.3 Detail Design with CAD Modeling

Detail design is then proceeded using the conceptual design B based on the evaluation matrix. The CAD model was modeled using SolidWorks. The CAD model includes the main building material of aluminum profile 2040 and pneumatic cylinders TN 16 × 75 mm. The model of the universal shape adapting jig after assembling is shown in Fig. 12.2. It consists of 12 pneumatic cylinders, 6 on each side to conform to workpiece shapes.

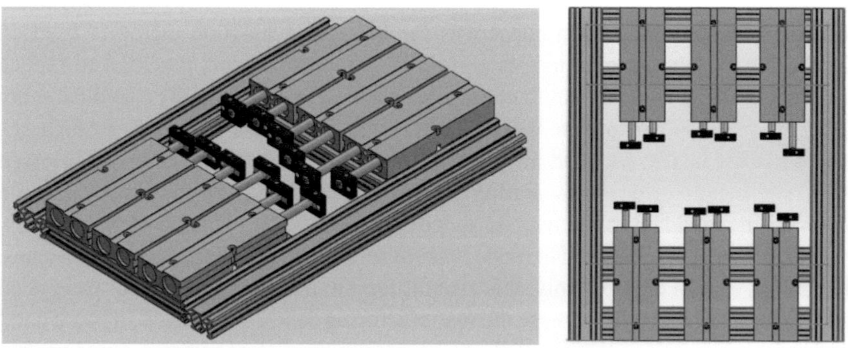

Fig. 12.2 CAD model of universal shape adapting jig (isometric and top view)

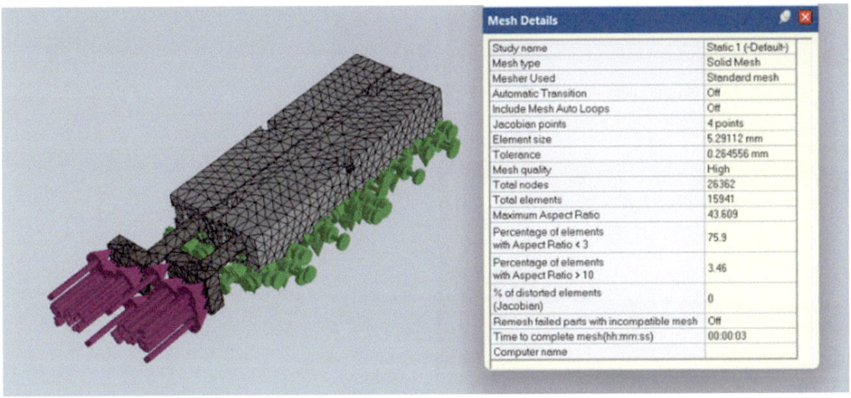

Fig. 12.3 Mesh details

12.3.4 Finite Element Analysis

Finite element analysis is done to determine the structural integrity of the jig under load. This method is used by engineers to simulate physical processes before testing with real prototypes. This helps to optimize the design process and optimize the components that will be fabricated. After creating a 3D model of the pneumatic cylinder, SolidWorks was used to do a static analysis to determine the circumstances under which the clamping process occurred. For the pneumatic cylinders base, the material was aluminum 6063-T5 and for the piston, the material was steel AISI 304.

The model was constrained at the bottom of the pneumatic cylinder since the pneumatic cylinder will be fixed to the aluminum profile base. The bottom of the pneumatic cylinder has been selected with fixed boundary condition in all axes because it will remain static during operation. The force for this analysis is tested at 1000 N. The entire mesh consisted of solid elements. The total number of elements of this mesh is 15,941 and the total number of nodes is 26,362 see Fig. 12.3.

12.4 Result and Discussion

12.4.1 Result of FEA Analysis

From the results obtained, it can be concluded that the pneumatic cylinder is safe to use with the clamping force of more than 1000 N. The maximum stress, if a clamping force of 1000 N is used, is only 67 MPa which is well below the tensile strength of AISI 304 steel at about 207 MPa. This does not affect the pneumatic cylinder because the result of the analysis did not exceed the yield strength, hence, no permanent deformation should occur (Fig. 12.4).

Fig. 12.4 Von Mises stress at 1000 N

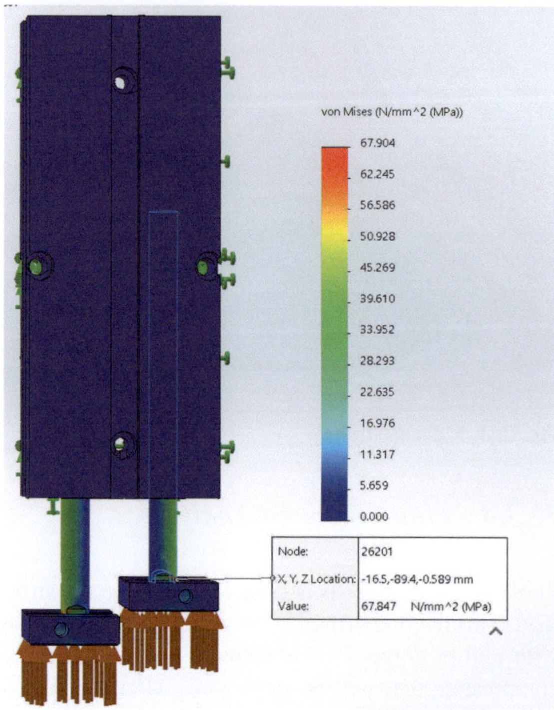

12.4.2 Prototype Fabrication and Performance Testing

Prototype fabrication is done and assembled properly, including all the necessary components, such as pneumatic tubing. The prototype consists of 12 pneumatic cylinders, multiple aluminum profiles with connectors and a pressure regulator as shown in Fig. 12.5. Testing is then done with multiple air pressure from 0.1 to 0.8 MPa. The clamping force is also measured by a push gauge.

From the testing data gathered in Table 12.2, a graph is plotted in Fig. 12.6. From the observation, it seems to indicate that there is a linear relationship between pressure in MPa and force in Newton. Specifically, it appears that the force is increasing by 5 Newtons for every 0.1 MPa increase in pressure.

This can be mathematically represented as:

$$\text{Force (N)} = 50 * \text{Pressure (MPa)} \tag{12.1}$$

This data could represent a physical system where the pressure is being applied uniformly over a certain area, such as this jig a piston pushing against a surface, and the force is being measured on the other side of the surface. By seeing this data, we can predict the force that will be generated by the cylinder when a specific pressure is applied. For example, if the jig needs to exert a force of 30 N, we know that we

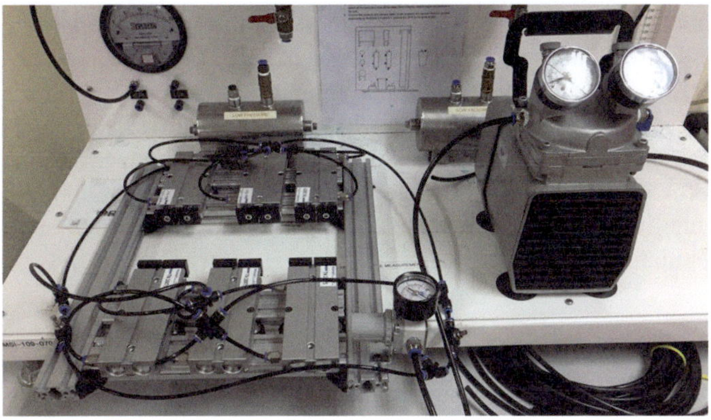

Fig. 12.5 Final model of the universal shape adapting jig

Table 12.2 Data for clamping test by controlling pressure

Pressure (MPa)	Force (N) per cylinder	Total force (N) per side (6 cylinder)
0.1	5	30
0.2	10	60
0.3	15	90
0.4	20	120
0.5	25	150
0.6	30	180
0.7	35	210
0.8	40	240

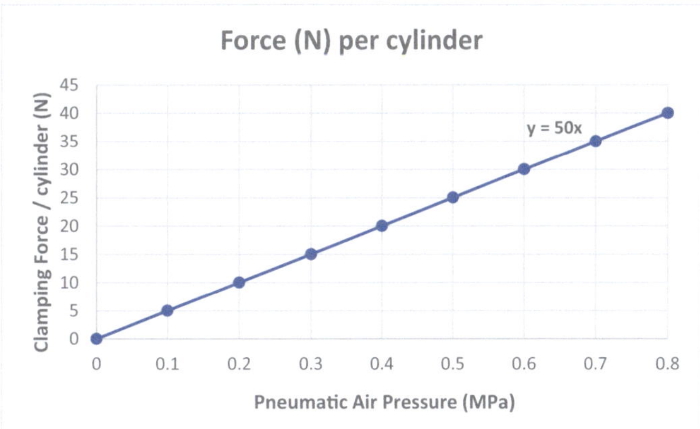

Fig. 12.6 Force versus pressure

need to apply a pressure of 0.6 MPa. In addition, the linear relationship between pressure and force allows to adjust the force generated by the cylinder by adjusting the pressure, which can be useful in fine-tuning the jig's performance to match the specific types of material that will use a low force to clamp. In summary, the data provided can be used to predict the force generated by a pneumatic cylinder, which can be useful in designing and developing a universal shape adapting jig. It also allows to fine-tune the performance of the jig by adjusting the pressure using the pressure regulator.

12.5 Conclusions

The design and fabrication project of a universal shape adapting jig was successfully done. It involves a combination of engineering design, analysis, testing and fabrication to optimize its performance during operation. The jig should be able to clamp a variety of shapes of workpieces via pneumatic pressure supplied, which was translated to force via the actuators. In summary, a universal adapting jig is essential because this jig can help to reduce production time and manufacturing cost of CNC processes.

References

1. Chikwendu Okpala C (2015) The design and need for jigs and fixtures in manufacturing. Sci Res 3:213
2. Flores Téllez G, Torres Méndez SJ (2006) Automated design of jigs: an application in CNC machine processes. In: Presented at the international symposium on robotics and automation, San Miguel Regla Hidalgo, México, August 25–28, ISRA code: ISRA LT17-control/automation
3. Chowdary BV, Richards MA, Gokool T (2018) Virtual conceptual design of a multi-purpose fixture for a CNC milling machine using the controlled convergence technique. West Indian J Eng 40:4–9

Chapter 13
Flow-Rotational Dynamics Interaction Effect on the Turbocharger System Performance

Khairul Shahril Shaffee, Siti Nor Nadirah binti Baba, Mior Firdaus bin Mior Abd Majid, Mohd Suyerdi Omar, Muhammad Najib bin Abdul Hamid, and Shahril Nizam bin Mohamed Soid

Abstract The turbocharger exhaust manifold is a channel that drives up the exhaust from the engine and plays an important role in the exhaust manifold. The exhaust manifold is installed at the engine due to heat and volumetric flow of the exhaust gases from the engine. The invention of the turbocharger needs to be advanced since the automotive industry keeps growing to make sure the performance of the turbocharger is at its best. There are several designs of exhaust manifolds that have been tested in this study to compare with the performance of the turbocharger. In this study, the method used was computational fluid dynamics to evaluate and compare the effect of three shapes of the exhaust manifolds on certain highlighted parameters. The focus was on validating certain chosen performance parameters such as velocity of air intake, pressure ratio of air, and rpm of the turbo wheel with experimental results. The result concluded which design was the best and having a great performance that was used commonly in the automotive industry.

K. S. Shaffee (✉) · M. F. bin Mior Abd Majid · M. S. Omar · M. N. bin Abdul Hamid · S. N. bin Mohamed Soid
Universiti Kuala Lumpur Malaysian Spanish Institute, Kulim Hi-Tech Park, 09000 Kulim, Kedah, Malaysia
e-mail: khairuls@unikl.edu.my

M. F. bin Mior Abd Majid
e-mail: miorfirdaus@unikl.edu.my

M. S. Omar
e-mail: msuyerdi@unikl.edu.my

M. N. bin Abdul Hamid
e-mail: mnajib@unikl.edu.my

S. N. bin Mohamed Soid
e-mail: shahrilnizam@unikl.edu.my

S. N. N. binti Baba
Centre for Women Advancement and Leadership, Universiti Kuala Lumpur, 50250 Kuala Lumpur Wilayah Persekutuan, Malaysia

T. A. A. Razak et al. (eds.), *Progress in Engineering Technology VI*, Advanced Structured Materials 229, https://doi.org/10.1007/978-3-031-81517-1_13

Keywords Exhaust manifold · Airflow analysis · Computational-aided design · Turbocharger

13.1 Introduction

A turbo manifold is a series of runners, or pipes that collect exiting exhaust gases which can be channeled into a merge collector. A flange is needed to bolt the turbo to the manifold and welded to the merge collector. There are four key components of the exhaust manifold, which are runner volume and size, collector angle toward turbo flange, runner length, and waste gate placement. This study aimed to improvise the performance of turbocharger exhaust manifold by doing an experiment and computational fluid dynamics (CFD) of the turbocharger exhaust manifold. The data from the experiments and from CFD simulation were taken. The manifold is a pipe or a chamber branching into several parameters. There are many types of manifolds, and they are dependent on many parameters [1]. This study was to compare three types of manifolds and the effect of the manifold design on turbo performance using cold blow.

13.2 Literature Review

13.2.1 Exhaust Manifold Application

The exhaust manifold is part of an engine component that can collect exhaust gases from the cylinder head and send them to the exhaust system [2]. The exhaust manifold plays an important task in the overall efficiency of a motor system [3, 4]. In many ways, the gas exchange cycle can be influenced by the exhaust manifold, such as piston work during the exhaust stroke, short-circuit fresh charge from intake into exhaust, and even cylinder filling [1].

The engine exhaust back pressure is characterized as the exhaust gas pressure created by the engine to exceed the exhaust system's hydraulic resistance to discharge the gases into the atmosphere. The exhaust back pressure is the gage pressure in the exhaust system at the exhaust turbine outlet in turbocharged engines, or the pressure at the exhaust manifold outlet in naturally aspired engines. The gas flow was regulated by a pressure gradient with the only possible flow path being that from a higher to a lower pressure. The gas cannot flow against increasing pressure. The engine pumps gas to a sufficiently high pressure by compressing it to overcome the flow obstructions in the exhaust system [5].

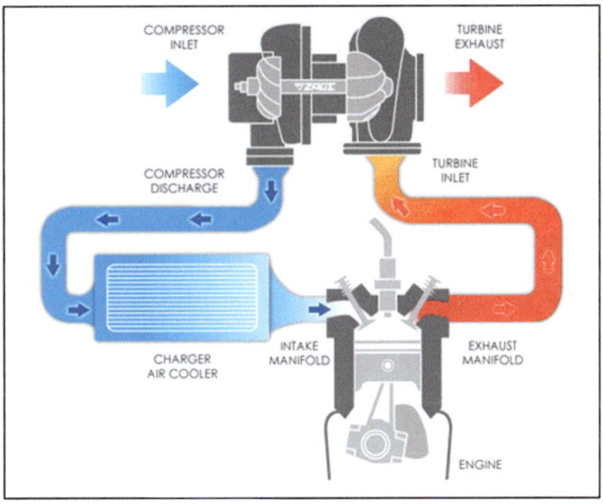

Fig. 13.1 Turbocharger system

13.2.2 Factors that Affect Turbo Performance on Manifold Design

The engine's exhaust gas uses the exhaust manifold to pipe out from the engine block to the turbocharger outlet (Fig. 13.1). The turbocharger's performance can be influenced by some factors from the manifold outlet, which are the radius, length, and geometry of the turbocharger. Other than that, the factors of turbo performance are the airflow velocity, turbo wheel rpm, and airflow pressure. These entire factors therefore need to be studied and developed so that the turbocharger has the best possible performance [6].

13.2.3 Method of Studying Turbo Performance

The exhaust manifold architecture has been developed for many years. Internal combustion engines' combustion efficiency can be enhanced by improving the exhaust manifold design. In early times, designs, which did not lead to a smooth flow, were quite common. The back pressure built up was much greater, which improved the work performed at the exhaust stroke by the specific piston. Significant quantities of residual gases remain in the compression chamber and the temperature increases as a result. At times, the manifold glows red-hot when working hard, requiring the use of asbestos and a highly heat-resistant fibrous silicate mineral, to protect the paintwork. Today, the exhaust manifold designs have been fully transformed. To develop earlier configurations, designers and researchers have come up with various designs involving various parameters affecting the exhaust manifold. Many studies have been reported on the impact of various parameters on the exhaust manifold analysis, design, and efficiency [7]. Traditional manifold optimization was based on the exhaust manifold processing. This form of trial and error can be successful but is too expensive and time-consuming. Besides this approach, the actual flow structure inside the manifold cannot be provided with any details. Using 3D CFD analysis, this vital knowledge can be obtained [8]. Ahmed et al. [9] proposed a comparison of five different designs of the exhaust manifold using the pressure and the velocity contour to determine the best possible model for the high velocity and the lower back pressure of the exhaust. The study was conducted by using CFD software. The overall energy losses in compressible flow at the compact exhaust manifold connection are calculated using CFD software [8, 10]

13.3 Methodology/Experimental Setup/Model Setup

13.3.1 Experimental Setup

The experiments were set up by assembling all the components. Then, run the experiments and check if there are any errors or if there are any leaks. After conducting the experiments, the data was recorded and the experiments were repeated by changing the manifold. Lastly, assembly of the component was done in the Solidworks software. The airflow simulation was conducted by using the three different types of manifolds (manifold A, manifold B, and manifold C) and getting the data from simulation (Figs. 13.2, 13.3 and 13.4).

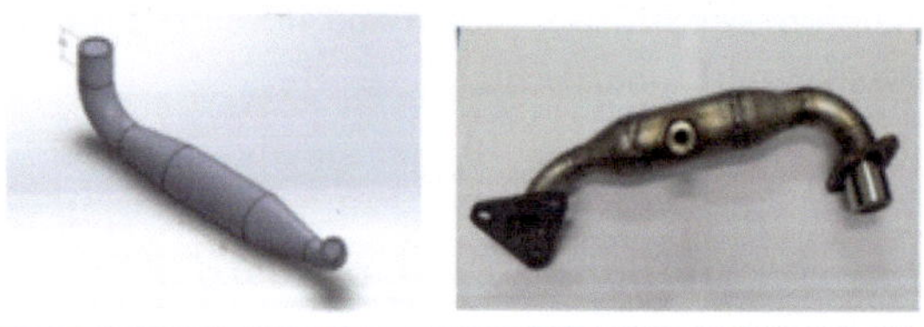

Fig. 13.2 Manifold A

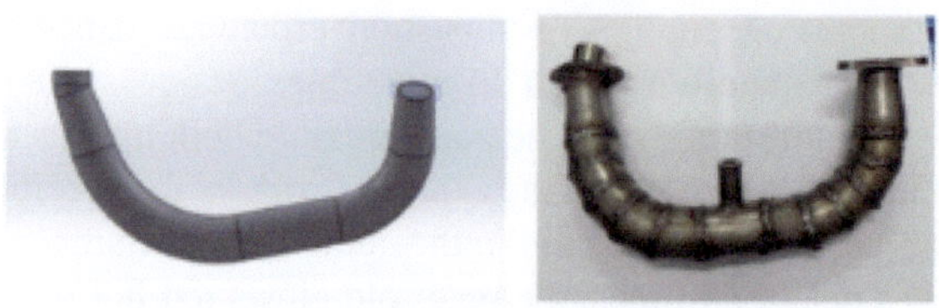

Fig. 13.3 Manifold B

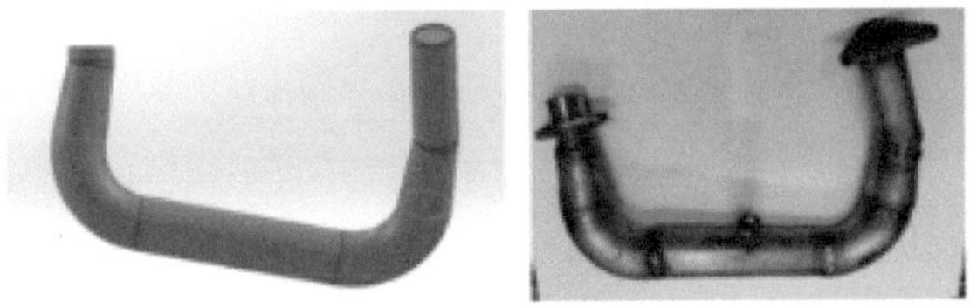

Fig. 13.4 Manifold C

13.3.2 Flow Simulation Process

In this study, the Solidworks software was used to analyze the data from flow simulation of three designs of the manifold. From the simulation of exhaust manifold, the airflow behavior inside the exhaust manifold can be observed clearly. Besides, the Solidworks software data can assist in validation of data from experiments to determine the air velocity and pressure. For this process, the designs of three different designs of the exhaust manifold have been made from previous study and literature. Considering the data collected from the simulations, the ideas and the method from the literature review have been applied. Then, the procedure of the simulation process was explained in detail in this part.

13.3.3 Data Collection

The experiment was carried out at the lab in UniKL MSI. Data was acquired from the reading of measuring devices such as anemometer, accelerometer, and pressure gauge. The reading was collected while the airflows in the exhaust manifold. The readings of rpm were collected while the turbine was running. Before running the experiments and collecting the data, leakage along the exhaust pipe was checked. In experiments of the turbocharger exhaust manifold, the position of the result being obtained was based on Fig. 13.5.

13.4 Result and Discussion

13.4.1 Velocity Inlet, Vin (Turbo)

The air velocity traveled from the inlet to the outlet of the exhaust manifold, and then into the turbocharger. Figure 13.6 shows the comparison of the inlet velocity for three manifold designs. Design A had a higher value of inlet velocity. To summarize that, manifold design A was chosen as the best design.

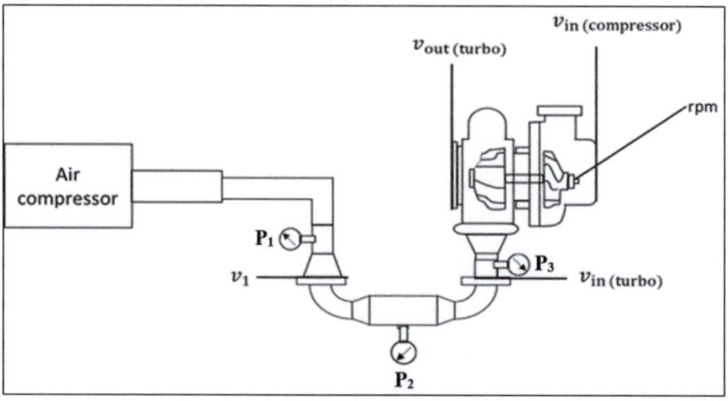

Fig. 13.5 Position to obtain data

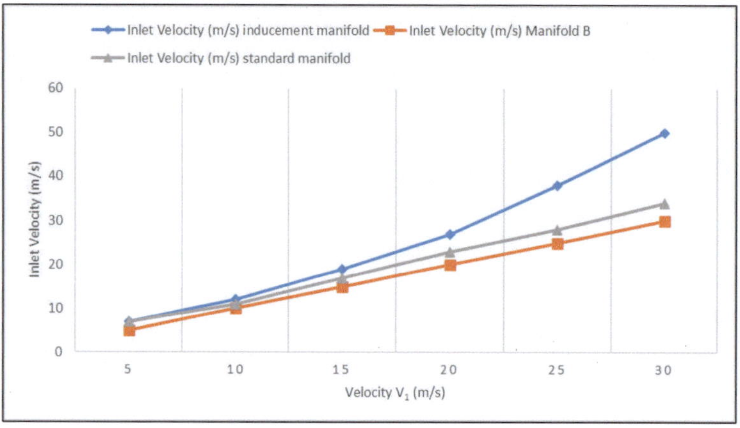

Fig. 13.6 Graph of velocity inlet

13.4.2 Exhaust Velocity, Vout (Turbo)

The highest value of the exhaust inlet velocity was for manifold design A. Figure 13.7 shows the comparison of exhaust inlet velocity for three manifold designs.

13.4.3 Compressor Inlet Velocity, Vin (Comp)

In terms of the compressor inlet velocity, the manifold design A had a higher value. The result shows the comparison value, and the worst was the manifold with design B. Figure 13.8 shows the comparison of the compressor inlet velocity between the three manifold designs.

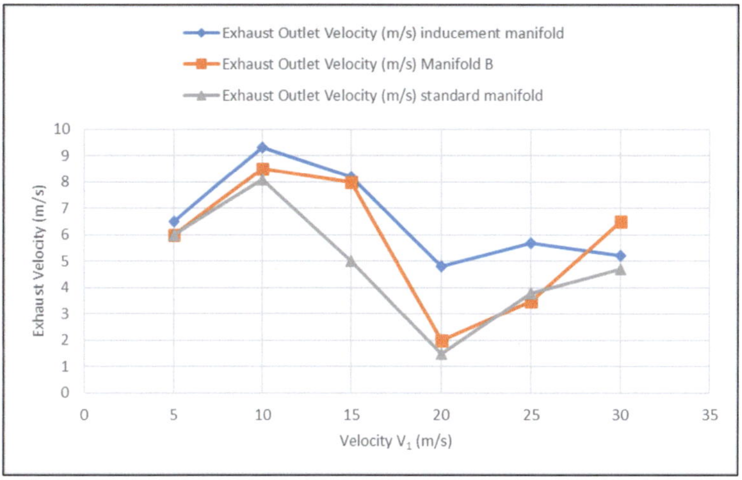

Fig. 13.7 Graph of exhaust velocity inlet

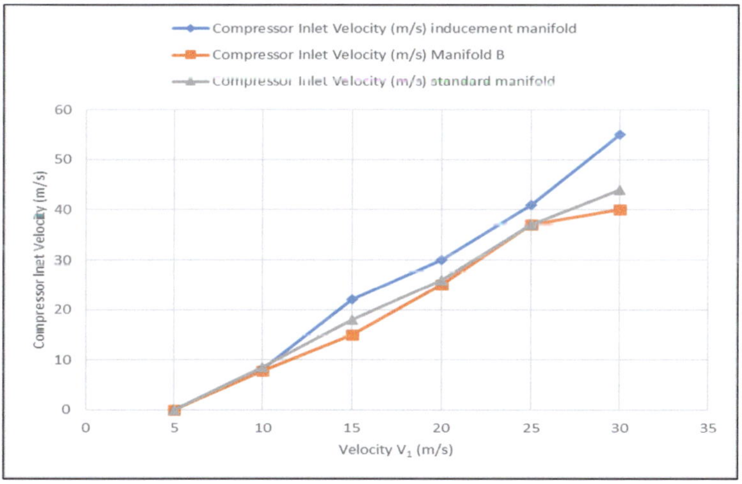

Fig. 13.8 Graph of compressor velocity inlet

13.4.4 Average Revolution Per Minute, rpm

The average value was calculated since three trials were conducted for one manifold design. From the observation, manifold A has higher values among their designs. It means that manifold A was the best design because of the higher average revolution per minute. Figure 13.9 shows a graph of average revolution per minute for three manifold designs.

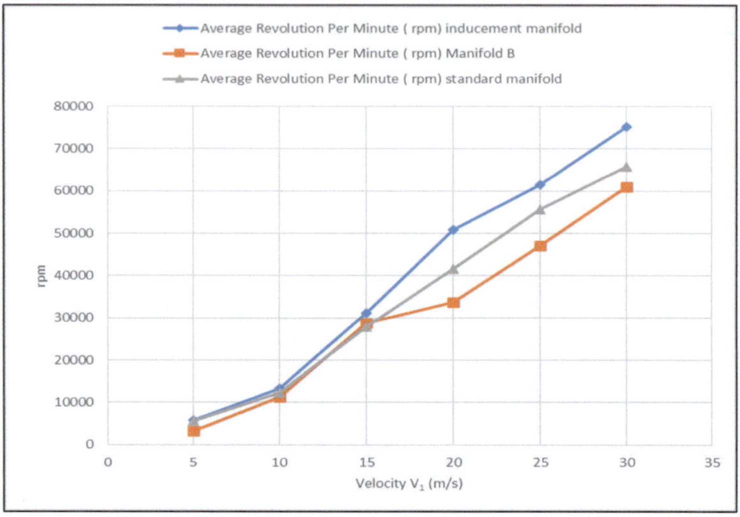

Fig. 13.9 Graph of average revolution per minute

13.5 Conclusion

The three manifolds were tested for each experiment and simulation. Manifold A had shown the best in most of the tests except for the fuel consumption test. In the simulations test, the turbocharger exhaust manifold A had a good performance compared to the turbocharger exhaust manifold B and turbocharger exhaust manifold C. At 30 m/s, the inlet velocity, Vin manifold A was 52 m/s while the manifold B and manifold C, which were 30 and 32 m/s. At 30 m/s inlet velocity, the outlet velocity produced from manifold A was 52 m/s, following by manifold B and manifold C, which were 30 m/s and 32 m/s. Manifold A had a good inlet velocity for the turbocharger. From the experimental tests, average revolutions per minute (rpm) at the turbocharger shaft using manifold A were the highest when velocity, V1, was 30 m/s which was 75,264 rpm compared with the turbocharger shaft using the manifold B and the manifold C, which were 60,957 and 65,774 rpm. Then, the compressor inlet velocity using manifold A also was the highest due to the highest rpm of turbocharger shaft which was 55 m/s compared to manifold B and manifold C which were 40 and 44 m/s. Design A manifold had a higher value for all the variables. The design B was the worst among three designs.

Acknowledgements The authors thank the Ministry of Higher Education (MOHE), Malaysia for the financial support provided via Fundamental Research Grant Scheme [ref no FRGS/1/2019/TK03/UNIKL/01/2]. The authors also thank Universiti Kuala Lumpur, Malaysia for the support provided via UniKL-STRG Grant [ref no str17042], and UniKL Malaysian Spanish Institute for providing the necessary facilities and resources to complete this study for publication.

References

1. Bajpai K, Chandrakar A, Agrawal A, Shekhar S (2017) CFD analysis of exhaust manifold of SI engine and comparison of back pressure using alternative fuels. IOSR J Mech Civ Eng 14(01):23–29
2. Kanawade N, Siras O (2016) A literature review on exhaust manifold design. Int J Sci Res Eng Tech (IJSRET) 5(5):286–289
3. Sunny Manohar D, Krishnaraj J (2018) Modeling and analysis of exhaust manifold using CFD. In: IOP conference series: materials science and engineering, vol 455, no 1, p 012132. IOP Publishing
4. Umesh KS, Rajagopal VPK (2013) CFD analysis of exhaust manifold of multi-cylinder Si engine to determine optimal geometry for reducing emissions. Int J Auto Eng Res Dev 45–56
5. Hushim MF, Alimin AJ, Razali MA, Mohammed AN, Sapit A, Carvajal JCM (2016) Air flow behaviour on different intake manifold angles for small 4-stroke PFI retrofit kit system. ARPN J Eng Appl Sci 11(12):7565–7571
6. Naeimi H, Domiry GD, Gorji M, Javadirad G, Keshavarz M (2011) A parametric design of compact exhaust manifold junction in heavy duty diesel engine using CFD. Therm Sci 15(4):1023–1033
7. Desantes JM, Galindo J, Guardiola C, Dolz V (2010) Air mass flow estimation in turbocharged diesel engines from in-cylinder pressure measurement. Exp Therm Fluid Sci 34(1):37–47
8. Whitley K (2010) Investigation of superturbocharger performance improvements through steady state engine simulation (Doctoral dissertation, Colorado State University)
9. Ahmed MS, Kailash, Gowreesh BA (2015) Design and analysis of a multi-cylinder four stroke SI engine exhaust manifold using CFD technique. Int Res J Eng Tech (IRJET) 2(9):927–938
10. Krishnara J, Rajesh Ruban S, Subramani N (2018) Analysis of exhaust manifold to improve the engine performance. Intl J Eng Tech 7(2):539–542

Chapter 14
A Comparative Study of the Velocity Profiles in Automotive Exhaust Manifolds

Khairul Shahril Shaffee, Aisyah Nasuha binti Mohd Noor, Mohd Suyerdi Omar, Khairul Akmal Shamsuddin, Shahril Nizam bin Mohamed Soid, and Ishak bin Abd. Azid

Abstract The exhaust system is one of the most important systems in automotive engines and functions to collect the exhaust gases from the cylinders, remove dangerous elements, reduce the degree of noise, and release the purified exhaust gases away from the vehicle's occupants at an appropriate location. The paper presents a study on the parameter and design of the manifold toward the performance of the engine and its improvement of standard manifolds. This study begins with the development of design and modeling of the manifold. The drafting and sketching have been carried out based on the parameters produced by the "Gap Study" method using a literature review and generated through the Taguchi method, Minitab. The process continued with designing the manifold by using CAD through Solidworks applications and finally produced a schematic technical drawing for preparation during the second stages. The simulation has been conducted through all the design stages and the result analyzed the velocity profile of the inlet, middle, and outlet. A comparison was conducted to find out the best manifold between the three models.

K. S. Shaffee (✉) · M. S. Omar · K. A. Shamsuddin · S. N. bin Mohamed Soid · I. bin Abd. Azid
Universiti Kuala Lumpur Malaysian Spanish Institute, Kulim Hi-Tech Park, 09000 Kulim, Kedah, Malaysia
e-mail: khairuls@unikl.edu.my

M. S. Omar
e-mail: msuyerdi@unikl.edu.my

K. A. Shamsuddin
e-mail: khairulakmal@unikl.edu.my

S. N. bin Mohamed Soid
e-mail: shahrilnizam@unikl.edu.my

I. bin Abd. Azid
e-mail: ishak.abdulazid@unikl.edu.my

A. N. binti Mohd Noor
Centre for Women Advancement and Leadership, Universiti Kuala Lumpur, 50250 Kuala Lumpur, Wilayah Persekutuan, Malaysia

T. A. A. Razak et al. (eds.), *Progress in Engineering Technology VI*, Advanced Structured Materials 229, https://doi.org/10.1007/978-3-031-81517-1_14

Keywords Exhaust manifold · CFD simulation · Velocity profile · Computer-aided software

14.1 Introduction

The exhaust system collects the exhaust gases from the cylinders, removes dangerous elements, reduces the degree of noise, and releases the purified exhaust gases away from the vehicle's occupants at an appropriate location. The exhaust manifold becomes the first component in the exhaust system [1]. Depending on the engine, the exhaust system may consist of one or two channels or designs [2]. The flow resistance must be chosen so that the exhaust backpressure has minimal impact on the engine performance. To guarantee that the exhaust system performs precisely, it must be evaluated systematically and designed properly [3]. Therefore, its components must be coordinated by the design engineers with the vehicle and engine. The exhaust systems of today have little in common with the simple systems of the past. In modern automobiles, the front section consists of the following which are the exhaust manifolds, the purification system, and the connecting pipes together with a rear section with the silencer system and pipes [4]. The entire system of the exhaust is attached to the floor pan via flexible mounting components [5]. Experimental setups were studied by previous researchers by calculating the velocity outlet using anemometers [6]. The researcher assesses the performance of the engine exhaust manifold, which is an important aspect in engine performance [7]. The other research investigates the air flow behavior inside different intake manifold angles. Six angles of the intake manifold have been investigated using a CFX simulation tool and the result indicated 180° as the best option for the intake manifold angle due to better air flow behavior inside the intake manifold [8]. This study used a computational fluid dynamics (CFD) simulation tool to obtain results in the form of velocity profiles in every inlet, middle, and outlet of the manifold designed. The knowledge gained from the study was able to understand the relationship between the velocity which affected with performance in terms of the combustion engine. This knowledge can be used in the future as a reference for similar research and development [9].

14.2 Literature

This section focuses in detail on the method to find the best design of the manifold to maximize the performance of the engine. The other study's reviews were useful in identifying and considering the method to measure and record the data that were obtained from experiment and simulation to improve the performance of the engine. All the factors that affect the performance of the engine performance have been considered in the studies and research papers that have a similar concept were discussed in more detail in this paper.

Fig. 14.1 3D model of exhaust manifold

14.2.1 Design of the Manifold

The performance of the engine depends upon the design of the exhaust manifold [8]. This paper's aim is to analyze the performance of the engine exhaust manifold. In this work, the 3D manifold design (Fig. 14.1) is prepared with the help of CAD software, and it is analyzed by the finite element software ANSYS. This CFD and thermal analysis was also done to check the performance of the redesigned exhaust manifold. The aim of CFD simulations is to investigate the volumetric efficiency behavior of an exhaust.

14.2.2 Experiment of the Exhaust Manifold

The comparison of several types of manifolds and the effect of the manifold design on the turbo performance were done using cold blow (Fig. 14.2). The result between the assembly of the manifold and the turbocharger was obtained in the study. The reading was taken by measuring the velocity, rpm, and pressure of the speed of the air and the speed of the turbocharger wheel. The result concluded which design was the best and has the best performance that was used commonly in the automotive industry [6]. There are four key components of the exhaust manifold, which are the runner volume and size, collector angle toward turbo flange, runner length, and waste gate placement.

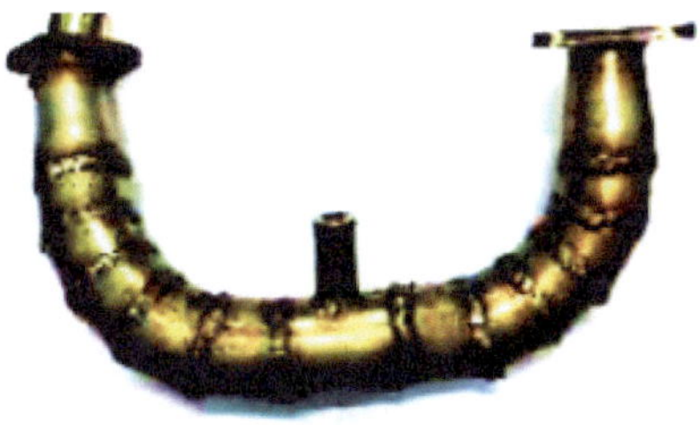

Fig. 14.2 Manifold design

All the equipment was set up and several designs of exhaust manifold that were involved in this study were introduced. The air compressor controller was used to control the velocity of the air. The turbo and manifold were assembled to get the velocity turbo compressor, the velocity exhaust, and the rotation per minute (rpm) of the turbine shaft for the three different manifold designs. The leaking of the air flow through the turbocharger exhaust manifold was checked [9].

14.3 Methodology

14.3.1 Simulation Setup

The simulation by using CFD method was run by the Solidworks application in the flow simulation tools. It is compulsory to complete the design before running the flow analysis by using computational fluid dynamics. The first step should be the sketching and drafting before continuing to design in the CAD software. Next, after designing, the model must go through the simulation process. To set up the simulation process, it is compulsory to set the "wizard" into the flow simulation. In the "wizard", there is an option to choose the analysis type, including internal and external flow. In this case, it is set as an internal flow and the fluids were chosen as air (gases). Additionally, the wall is set to be adiabatic, with the default environmental temperature of 293.2 Kelvin.

14.3.2 Boundary Conditions

An important part of any mathematical model is its boundary conditions. The function is to guide the flow. Common boundary conditions in ANSYS FLUENT or Solidworks Flow (and other CFD programs) include inlets, outlets, walls, and symmetrical regions. For this study, the boundary conditions are set out in two different LIDs. LID must be created before continuing with the flow simulation and it was quite challenging, this is due to the selection of the faces to choose the lid out of hundreds of choices. For boundary conditions, LID number 1 is selected as inlet velocity. The purpose of this inlet velocity is to insert a certain amount of velocity. The other boundary conditions were environment pressure, which is located on the outlet of the manifold, thus, LID 2 was chosen in this case.

14.4 Result and Discussion

Before starting the simulation process, it is compulsory to go through a grid-independent test. This is to ensure the best meshing grid for the entire simulation. After the grid-independent study was completed, it is obligatory to do a validation from the literature review. The validation must not exceed 10% difference to make sure the result was valid and genuine. The process of the simulation continued with the result of inlet velocity and outlet velocity of manifold A, manifold B, and manifold C (Figs. 14.3, 14.4, and 14.5).

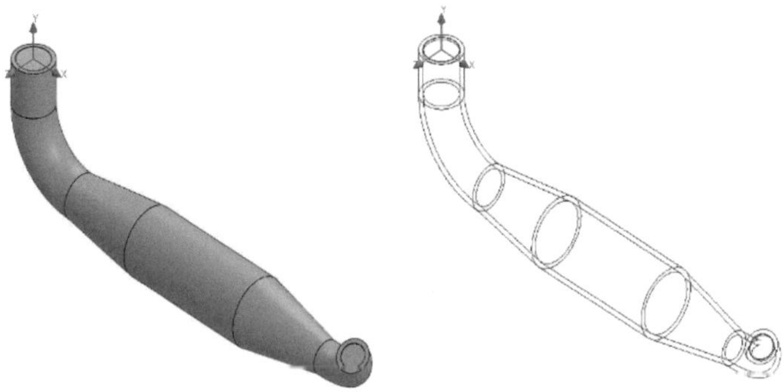

Fig. 14.3 CAD design of manifold A (shaded with edges and wireframe display style)

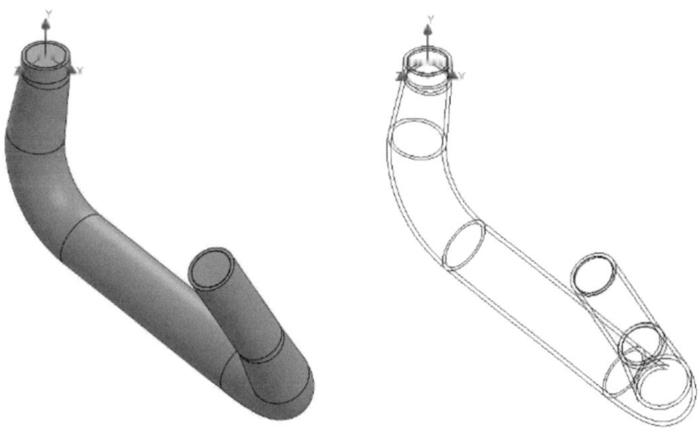

Fig. 14.4 CAD design of manifold B (shaded with edges and wireframe display style)

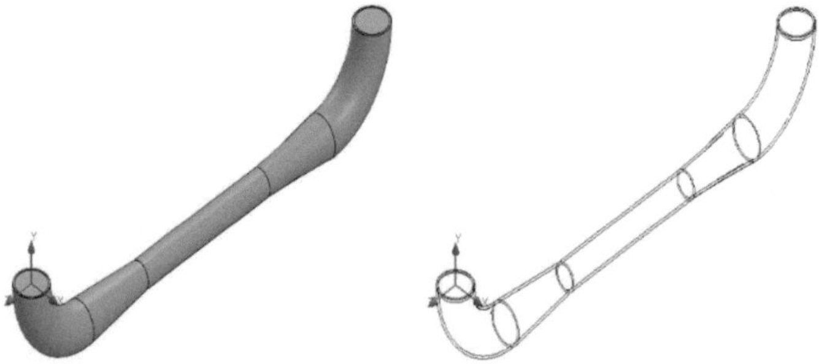

Fig. 14.5 CAD design of manifold C (shaded with edges and wireframe display style)

14.4.1 *Drafting and Sketching*

This result focuses on rough drafting and sketching. Sketching is an engineering method for rapidly exploring ideas. One or two hours of sketching can be used to generate numerous potential answers to a design problem. This is a crucial phase in the design process. Although it is feasible to create sketches on a computer, it is slower than sketching several ideas on paper.

14.4.2 *CAD Designing*

This result focuses on the CAD drawing and modeling. CAD drawing or modeling, often known as computer-aided design, is a crucial component of the design process. Computer-aided design is utilized for a range of purposes, from 3D printing prototypes to photorealistic promotional graphics.

14.4.3 *Simulation Results*

The simulation results include each manifold inlet and outlet of the velocity profile since the objective of this study is to compare the velocity profile of each segment of every model of manifold. The inlet velocity and outlet velocity are presented in Fig. 14.6. Finally, the best model was determining the velocity parameter.

The outlet velocity produced by every model of manifold with selected value of the inlet velocity ranges from 5 to 30 m/s. The maximum outlet velocity was selected as the largest value from the simulation process. Manifold A produces the best outlet velocity of all three models tested. The second best result falls to manifold C and the last one is manifold B, which is the standard manifold. At the 30 m/s of inlet manifold velocity, which is the highest result obtained, manifold A produces about 55.02% different to manifold C with the value 70.09 m/s toward 30.82 m/s on manifold C.

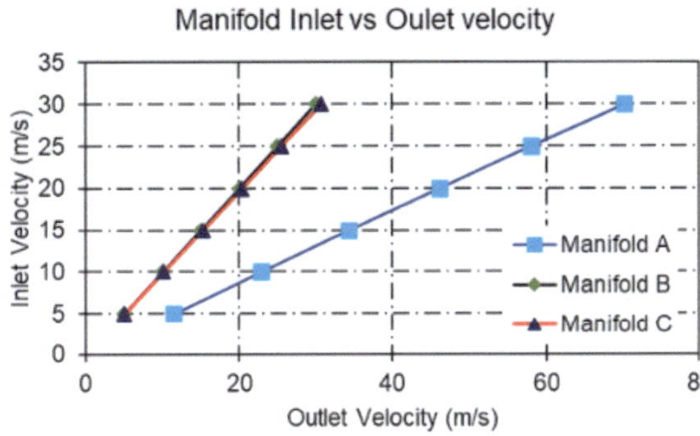

Fig. 14.6 Graph inlet vs outlet velocity of all manifolds

The result showed that the velocity of outlet increased with the increase of inlet velocity at every manifold. In other words, the larger the inlet velocity, the more the outlet velocity produced by all models. The data shows that manifold A was the best among all the models in terms of velocity outlet. For additional information, Fig. 14.7 shows the illustration of velocity cut-plot with vector of manifold A at 30 m/s and Fig. 14.8 shows the velocity flow trajectories of manifold A at 30 m/s.

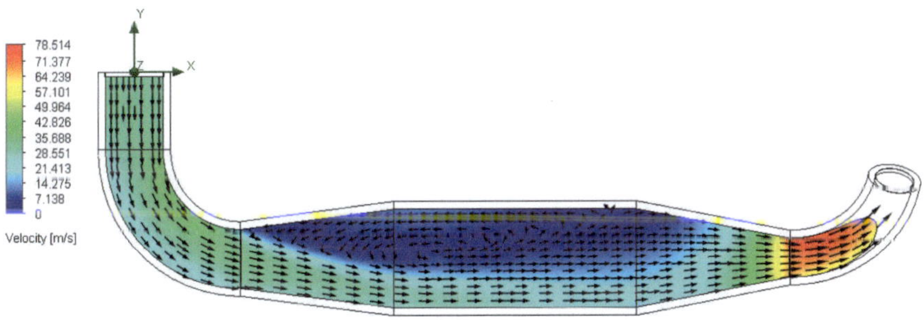

Fig. 14.7 Velocity cut-plot with vector of manifold A at 30 m/s

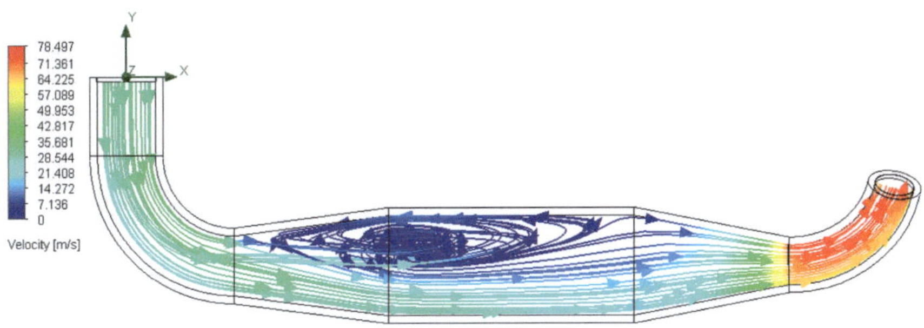

Fig. 14.8 Velocity flow trajectories of manifold A at 30 m/s

14.5 Conclusions

The relationship between each design parameter toward the performance of the engine is clearly identified through the parametric study. The generated parameters were used to complete design process and finally, the simulation to obtain data for creating the velocity profile and the comparison were done among the manifold models.

Acknowledgements The authors thank the Ministry of Higher Education (MOHE), Malaysia for the financial support provided via Fundamental Research Grant Scheme [ref no FRGS/1/2019/TK03/UNIKL/01/2]. The authors also thank Universiti Kuala Lumpur, Malaysia for the support provided via UniKL-STRG Grant [ref no str17042], and UniKL Malaysian Spanish Institute for providing necessary facilities and resources to complete this study for publication.

References

1. Bajpai K, Chandrakar A, Agrawal A, Shekhar S (2017) CFD analysis of exhaust manifold of SI engine and comparison of back pressure using alternative fuels. IOSR J Mech Civ Eng 14(01):23–29
2. Chaudhri JH, Patel BS, Shah SA (2014) Muffler design for automotive exhaust noise attenuation-a review. J Eng Res App 4(1):220–223
3. Cihan Ö, Bulut M (2019) CFD analysis of exhaust manifold for different designs. Euro Mech Sci 3(4):147–152
4. Hushim MF, Alimin AJ, Razali MA, Mohammed AN, Sapit A, Carvajal JCM (2016) Air flow behaviour on different intake manifold angles for small 4-stroke PFI retrofit KIT system. ARPN J Eng App Sci 11(12):7565–7571
5. Kammuang-Lue N, Boonjun J (2018) Design of intake manifold and selection of suitable material for intake manifold gasket for student formula. In: Proceedings of 2018 9th international conference on mechanical and aerospace engineering, ICMAE 201, pp 511–517
6. Kanazaki M, Morikawa M, Obayashi S, Nakahashi K (2003) Exhaust manifold design for a car engine based on engine cycle simulation. In: Parallel computational fluid dynamics 2002. North-Holland, pp 475–482
7. Krishnara JC, Rajesh Ruban S, Subramani N (2018) Analysis of exhaust manifold to improve the engine performance. Inter J Eng Tech 7(2.8):539
8. Ling N, Rangshu X, Wei X, Minli B (2011) CFD simulation calculation of inlet manifold modification design. In: 2011 international conference on electric information and control engineering. IEEE, pp 2035–2038
9. Raghuwanshi G, Kakirde A, Sharma S (2018) Design and analysis of exhaust manifold comparing different specifications. Int J Eng Trends Technol 62(1):42–45

Chapter 15
Taguchi Method's Investigation of Infill Settings Impact on Tensile Strength of 3D-Printed PLA Safety Hooks

Wan Muhammad Almahdi Wan Rasyed, Khairul Akmal Shamsuddin, Ahmad Kamal Ismail, Khairul Shfigahril Shaffee, and Mohd Faizal Abu Talib

Abstract This study aims to find a new lightweight and durable material for safety hook applications. The study requires experience in the necessary steps, starting from identifying the problem statement. The research objectives are to analyze the optimum infill settings for the PLA safety hook in achieving ultimate tensile strength via the Taguchi method. Finding out how to get the output and why to choose the strategy requires conducting a literature review. The ideal specifications for three-dimensional (3D) printers were a major focus of the literature review. Modeling the safety hook's design using actual dimensions as a guide is crucial. The Taguchi method is used to generate numerous combinations of parameters and ranges. It was then converted to a stereolithography file and printed on a 3D printer. Each sample is placed directly into the universal testing machine for a tensile test. Computer-aided design software simulation was used to record and verify the results. The goal of the data analysis was to contrast the printed samples' tensile strength characteristics with those of the pre-existing material used in the application. Correlating the best test results with the attributes of the already available material is essential. The data was subsequently confirmed, and the researchers concluded that PLA was suitable for use in safety hook applications utilizing an alternate fabrication method, i.e. 3D

W. M. A. W. Rasyed · K. A. Shamsuddin (✉) · A. K. Ismail · K. S. Shaffee · M. F. A. Talib
Universiti Kuala Lumpur, MalaysianSpanishInstituteKulimHi-TechPark, 09000 Kulim, Kedah, Malaysia
e-mail: khairulakmal@unikl.edu.my

W. M. A. W. Rasyed
e-mail: almahdi_rasyed@s.unikl.edu.my

A. K. Ismail
e-mail: ahmadkamal@unikl.edu.my

K. S. Shaffee
e-mail: khairuls@unikl.edu.my

M. F. A. Talib
e-mail: mfaizal@unikl.edu.my

T. A. A. Razak et al. (eds.), *Progress in Engineering Technology VI*, Advanced Structured Materials 229, https://doi.org/10.1007/978-3-031-81517-1_15

printing. After the validation process, formulating a determination is necessary as to whether all objectives were met or the opposite.

Keywords 3D printing · Safety hook · PLA · Infill · Taguchi · Stress · Strain · ASME B30.10 · CAD · UTM machine · SolidWorks

15.1 Introduction

The difficulty of lifting and mooring operations has prompted safety hooks to advance in sophistication, just like human evolution and history have done. Hooks were previously made of special single-grade carbon steel, and even if the working load increased, there was no need to change the material or design of the hook to make it larger. Alloy steels were the first steel grades that allowed for an increase in working load without requiring a change in size or weight. Nowadays, 3D printing is frequently used as an additive manufacturing (AM) technique due to its simplicity in producing complicated designs that are easily customizable and able to reduce waste made from any materials. Mechanical behavior is often a critical aspect of printed part usage that is hard to define. However, the mechanical properties depend on process parameters and filling patterns. Therefore, there has been a significant interest in studying the effect of infill settings on the mechanical properties of 3D-fabricated parts. Determining which infill density percentage, infill pattern types, and infill layer thickness provide the ultimate tensile strength for safety hook application is the actual goal.

Numerous investigations have been done to determine the characteristics of the PLA. The qualities of PLA must be good or equal to those of the present material used in the same application if it is to be used for safety hook applications. As a result, the focus of this research is on the characteristics of PLA, the sole material employed. The purpose of PLA was to create a revolutionary material for safety hook applications. Although the characteristics of this material may vary, the research will only pay attention to the mechanical characteristics, which are solely strength characteristics, and not dynamic characteristics. Polylactic acid, or PLA, is a typical thermoplastic polymer. The ease of manufacture and inexpensive production costs of this material contribute to its popularity. Gokhare stated that PLA is more environmentally friendly than other polymers because it is made of renewable resources and is biodegradable [1]. The biocompatibility of PLA with the human body is one of its other outstanding qualities. It also has a harder structural makeup than ABS and melts at a lower temperature than ABS, between 180 and 220 °C. The PLA's glass transition temperature ranges from 60 to 65 °C. These characteristics demonstrate that, in the modern world, PLA may be a viable solution for any project application. The mechanical characteristics of PLA mentioned in the study are shown in Table 15.1.

Previous investigation reveals that a higher percentage of infill and a greater number of solid contours or outer shells improve the ultimate tensile strength [2],

Table 15.1 Taguchi design data

	C1-T	C2-T	C3-T	C4	C5
	Infill pattern	Infill density (%)	Infill layer thickness (mm)	Tensile strength, Rm (N/mm^2)	Force, Fm (N)
1	ZigZag	40	0.12	2.03	841.5
2	ZigZag	60	0.16	2.14	885.0
3	ZigZag	80	0.20	3.16	1312.5
4	Lines	40	0.12	2.13	883.0
5	Lines	60	0.16	3.07	1274.5
6	Lines	80	0.20	3.26	1352.0
7	Grid	40	0.12	3.17	1314.5
8	Grid	60	0.16	4.25	1763.5
9	Grid	80	0.20	5.66	2345.5

whereas raising the infill density percentage improved the mechanical properties of the PLA [3]. The feed rate and infill patterns, on the other hand, have little effect on the mechanical properties. Furthermore, the ultimate tensile strength and yield strength of the samples were directly affected by the infill density percentage [4]. The proportion of infill density affects the modulus of elasticity. The infill density percentages in all PLA samples collapsed brittlely, with no obvious necking or distortion before failure. 3D-printed parts generated with a deposition angle of 90° for the rectangular infill pattern have better mechanical strength as the infill density increases [5]. Because increasing infill density reduces the size of the infill pattern, it has a significant impact on the overall printing time. As a result, to print the same part, the nozzle must travel a greater distance. Increasing the infill density increases the mechanical strength of the part, making it more solid and reducing the number of cavities [6]. The ultimate tensile strength decreases as the layer height increases. The ultimate tensile strength of a building that is upright and on the edge is reduced.

In another work, it is found that the specific tensile strength decreases as the infill density percentage decreases [7]. Furthermore, the specific tensile strength becomes slightly higher when the items are printed with hexagonal pattern infill geometry, making them stiffer than parts with rectilinear infill geometry. Parts with hexagonal infill demonstrated higher constant tensile strength as a function of orientation than those without. The 3D-printed part with a different type of infill pattern at the same infill density percentage and the same number of perimeter shells will create a contrast in stiffness and strength [8]. At the same density, the part that uses a higher ratio of perimeter shells for the same infill pattern enhances stiffness and strength. The latest study used four variables: printing speed, infill ratio, printing temperature, and layer thickness [9]. 3D printing is done at speeds of 20, 40, and 60 mm/s. The infill ratio was set at 20, 40, and 60%, and the printing temperature was set at 350 °C, 360, and 370 °C. A number between 0.20, 0.25, and 0.30 mm were the layer for the thickness. According to the research, a printing setup with a speed of 60 mm/s, a layer thickness of 0.20 mm, a printing temperature of 370 °C, and an infill ratio of 20% results in a maximum tensile strength of 40.0 ± 4.4 MPa.

This study focuses on applying PLA as a material for the application of a safety hook based on the ASME standards, volume B30.10, and the ISO 7597 standards due to its optimal design, or design-wise, that could withstand stress at a particular level before splitting into two parts. The major focus of this study is to identify the 3D printing settings needed to produce a high-strength PLA safety hook.

Since the output emphasizes the ultimate tensile strength of the printed product, simulated results and comparisons between the data from this study and the data from the previously used material are used to validate the data. This research can help determine whether PLA is appropriate for use in the safety hook application.

15.2 Methodology

The goal of a new material to be used in applications for safety hooks is the focus of this study. The material for this study was chosen to be PLA. Two factors played a role in the decision to use PLA: first, UniKL MSI's "Ender 3" 3D printer's limitations, which were taken into consideration. PLA was one of the materials that could be printed with the "Ender 3" 3D printer, but only under certain conditions. The fact that PLA is more dependable is another aspect that influenced the material choice. Along with being a little more environmentally friendly, it is also simpler to get and use. SolidWorks software was applied during the design phase. The shape and dimensions of the actual hook were similar to the shape and dimensions of a 9/32-inch ASME B30.10 safety hook shown.

15.2.1 Tensile Strength Test

The specimens were to be tested after printing. All nine specimens were chosen and put through the tensile test on the universal testing machine (UTM) to meet the research's output requirements. The UTM equipment employed was the "Instron 5582" universal testing machine, as depicted in Fig. 15.1. The reason for using this machine in the first place was that it was accessible at the Materials Laboratory, UniKL MSI. Moreover, Fig. 15.1 illustrates the end results of the tensile test done on the UTM machine for all nine specimens. It shows the same broken or failure area, which means every test was a success.

Fig. 15.1 Tensile test and outcomes

15.2.2 SolidWorks Simulation and Taguchi Approach

The simulation in SolidWorks was used to visualize the outcomes. To verify the tensile test results, simulated data was used. From there, the data was compared and confirmed. The original study's first goal was to figure out the 3D printing parameters for PLA safety hooks. Minitab software was used to enter the data into the Taguchi method after the parameters and levels had been defined. It was then possible to obtain a list of several sample parameters. The printing of the sample was done using the parameter combination created. The Taguchi analysis method was used to insert the data (Table 15.1) into Minitab once the outcomes were confirmed to determine the ideal printing parameters for a PLA safety hook.

15.3 Result and Discussion

From the collected data of the tensile test, it was found that the combination of a grid pattern, 80% infill density, and 0.20 mm infill layer thickness produced the highest tensile strength at a value of 5.66 MPa as shown in Fig. 15.2.

The values obtained from the simulation data (Figs. 15.3a, b, c) were nearly identical to the tensile tested values of 3D-printed safety hooks. The highest stress value of the tensile tested safety hook was 5.66 MPa, while the simulation obtained a stress value of 5.731 MPa. This is a 1% difference.

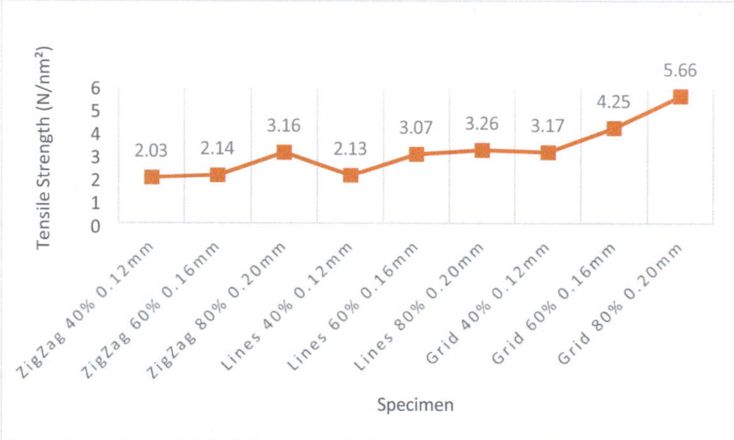

Fig. 15.2 Tensile strength data graph

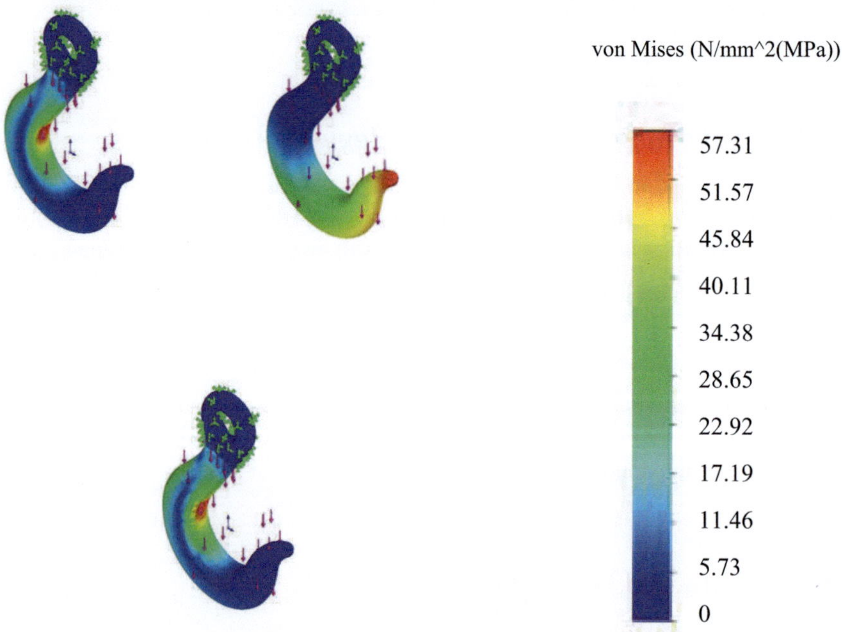

Fig. 15.3 **a** Stress result. **b** Displacement result. **c** Strain result

Table 15.2 Signal-to-ratio parameter ranking

Larger is better			
Level	Infill pattern	Infill density	Infill layer thickness
1	10.59	10.59	10.59
2	11.87	12.65	12.67
3	15.56	14.78	12.93
Delta	4.96	4.19	0.51
Rank	1	2	3

Table 15.3 Means parameter ranking

Level	Infill pattern	Infill density	Infill layer thickness
1	507.7	507.7	651.8
2	586.3	655.4	661.1
3	906.1	837.0	687.2
Delta	398.4	329.3	35.4
Rank	1	2	3

In Tables 15.2 and 15.3, Figs. 15.4, and 15.5, both the signal-to-noise ratio and mean calculations of the infill parameters show the same pattern of graph, which is in increasing order. Infill pattern was ranked first for the parameter that has an influence on the tensile strength property, whereas infill density was second, and infill layer thickness took last place. The infill pattern has a significant impact on the tensile strength of 3D- printed safety hooks. Therefore, the infill pattern can be considered the optimum infill parameter for tensile testing.

15.4 Conclusions

The main purpose of this experiment is to create a safety hook that is reinforced with polylactic acid as the material. Moreover, the effect of infill variations on the tensile properties of a 3D-printed PLA safety hook was investigated. The goal has been met with success. PLA has a high tensile strength property that can withstand heavy loads, which was proven by the SolidWorks simulation. The Taguchi analysis shows that the infill pattern has the highest influence on tensile properties other than infill density and infill layer thickness. Tensile strength was at its peak when accustomed to a grid pattern and an 80% infill density with an infill layer thickness of 0.20 mm. This research needed some optimization of the safety hook design to be more suitable for heavy-duty activities. Other than that, future researchers need to use a more durable plastic filament like an industrial grade filament (if cost is not a problem). If that is not possible, consider testing using the common ones such as ABS and PTEG.

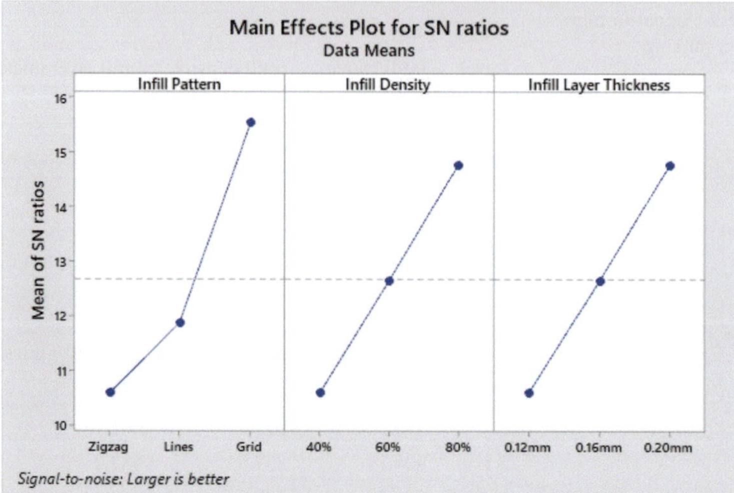

Fig. 15.4 Signal-to-noise ratio main effects graph

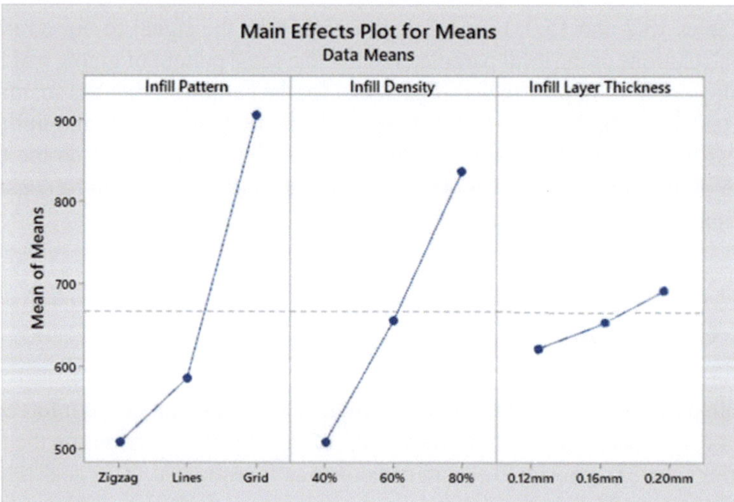

Fig. 15.5 Means main effects graph

References

1. Gokhare VG, Raut DN, Shinde DK (2017) A review paper on 3D-printing aspects and various processes used in the 3D-printing. Int J Eng Res Technol 6(06):953–958
2. Attoye S, Malekipour E, El-Mounayri H (2019) Correlation between process parameters and mechanical properties in parts printed by the fused deposition modeling process. In: Mechanics

of additive and advanced manufacturing, volume 8: proceedings of the 2018 annual conference on experimental and applied mechanics. Springer, pp 35–41

3. Qattawi A, Alrawi B, Guzman A (2017) Experimental optimization of fused deposition modelling processing parameters: a design-for-manufacturing approach. Procedia Manuf 10:791–803
4. Johnson GA, French JJ (2018) Evaluation of infill effect on mechanical properties of consumer 3D printing materials. Adv Technol Innov 3(4):179–184
5. Abbas T, Othman FM, Ali HB (2017) Effect of infill parameter on compression property in FDM Process. Dimensions 12(7):25–4
6. Rodríguez-Panes A, Claver J, Camacho AM (2018) The influence of manufacturing parameters on the mechanical behaviour of PLA and ABS pieces manufactured by FDM: a comparative analysis. Materials 11(8):1333
7. Farbman D, McCoy C (2016) Materials testing of 3D printed ABS and PLA samples to guide mechanical design. In: International manufacturing science and engineering conference, vol 49903. American Society of Mechanical Engineers, p V002T01A015
8. Lubombo C, Huneault MA (2018) Effect of infill patterns on the mechanical performance of lightweight 3D-printed cellular PLA parts. Mater Today Commun 17:214–228
9. Deng X, Zeng Z, Peng B, Yan S, Ke W (2018) Mechanical properties optimization of poly-ether-ether-ketone via fused deposition modeling. Materials 11(2):216

Chapter 16
Silent MHD Thruster for Micro Underwater Vehicle—Stealth Application

Hanis Syamira Binti Mohd Kamal, Mior Firdaus Mior Abd Majid, Johan Ihsan Mahmood, Ahmad Kamal Ismail, and Khairul Shahril Shaffee

Abstract Many underwater vehicles still use propellers to move their vehicle. The disadvantage of the traditional propeller is that it produces noise and vibration. Nowadays, with the technology that has been invented and improved, there is another way to move the underwater vehicle besides using the propeller which is the use of a magnetohydrodynamic (MHD) thruster. Additionally, the design reduces wear and friction throughout the drivetrain. When vibration and noise are reduced, it enhances passenger comfort. The magnetohydrodynamic thruster that is built in this paper is an open rectangular pipe that has aluminum electrodes attached to the wall of the pipe and neodymium magnets. As this experiment is for underwater vehicles, a salt solution was used to represent the seawater. A permanent magnet supplies the magnetic field in this circuit, and an electrode creates the flow's electric field. Fluid conductors travel throughout the circuit as a result of magnetohydrodynamic processes, the salt water exposed to the magnetic field and electric current. The cross product of a voltage supplied and the magnetic field applied by the permanent magnet produces the electromagnetic force. The main parameters of this experiment are magnetic strength, voltage, sound level, and volume of flow rate. The water flow rate is then measured in terms of volume. A comparison between conventional and MHD thruster is to validate that the MHD thruster can reduce noise and vibration.

H. S. B. M. Kamal
Centre for Women Advancement and Leadership, Universiti Kuala Lumpur, 50250 Kuala Lumpur, Wilayah Persekutuan, Malaysia
e-mail: hanis.kamal@s.unikl.edu.my

H. S. B. M. Kamal · M. F. M. A. Majid (✉) · J. I. Mahmood · A. K. Ismail · K. S. Shaffee
Universiti Kuala Lumpur, Malaysian Spanish Institute Kulim Hi-Tech Park, 09000 Kulim, Kedah, Malaysia
e-mail: miorfirdaus@unikl.edu.my

J. I. Mahmood
e-mail: johanihsan@unikl.edu.my

A. K. Ismail
e-mail: ahmadkamal@unikl.edu.my

K. S. Shaffee
e-mail: khairuls@unikl.edu.my

T. A. A. Razak et al. (eds.), *Progress in Engineering Technology VI*, Advanced Structured Materials 229, https://doi.org/10.1007/978-3-031-81517-1_16

It was found that the MHD thruster is more reliable compared with the conventional thruster. Nevertheless, there are still more works to do, to make the MHD thruster better.

Keywords Magnetohydrodynamic (MHD) · Noise · Vibration · Comfort · Neodymium magnets (NdFeB) · Seawater · Magnetic field · Electric current

16.1 Introduction

The study of conducting fluids interacting with electric and magnetic fields is known as magnetohydrodynamics, or MHD. Similar to a piston engine generator or a turbo-generator, the MHD generator converts the internal energy of gas into electric power, and the underlying physical phenomenon is the same in both cases. The energy of the gas is transformed into the motion of solid conductors in a magnetic field in the first two devices. The gas itself serves as the conductor of the MHD generator. A magnetic field is traversed by this conducting gas as it expands through the nozzle, producing an electromotive force and current flow in accordance with Faraday's equations of induction [1].

Ship propulsion systems come in many different varieties today. The majority of ships still rely mostly on fixed propellers for propulsion. The current propulsion system is extremely intricate, particularly in terms of its mechanical components. The mechanical system's vibration and noise issues would affect how comfortable the passengers are. Thus, in this work, the magnetohydrodynamic propulsion system is presented to overcome the vibration and noise difficulties [2].

Magnetohydrodynamic thrusters are used for underwater vehicles in the salt water that simulates seawater in this project. The magnetohydrodynamic thrusters are silent due to their no-moving-parts propulsion, which qualifies it for military use and allows them to be implemented in submarines and other military underwater vehicles. This has a connection to the stealth application. Submarines make up a sizable percentage of the navies of most superpowers and have been essential to naval conflict since World War I. They can operate covertly in international waters, unlike any other type of vessel. The submarine is a practical and efficient weapon because of stealth application.

Since MHD thrusters do not require the conventional rotational drive component, they have long been seen as viable options for ship propulsion. The main issue with this project is that the conventional thruster employed an outdated bladed propeller. Propeller blade rotation and noise are produced when the engine produces torque, making them unsuitable for stealth applications. The undersea vehicle moves more quietly while employing MHD thrusters since they combine magnetic and hydraulic energy.

To make the water flow effectively and smoothly so that the vehicle can move, the MHD thruster requires an incredibly strong magnet. The water will move less if the magnet is not strong enough. In addition, the quantity of magnets has an impact on how the water moves. The water will flow more quickly when more magnets are used.

Additionally, the liquid that will be used will influence the outcome based on its qualities. The speed of the water flow differs as a result of the various liquids that are used. The ability of a liquid to conduct electricity increases with density. Salt water is a good illustration. When compared to water, salt water is denser. As a result, the water flow adjusts as needed. However, the salt concentration needs to remain constant during the experiment in order to measure the water flow.

The objectives of this project are to set up the experiment with all the apparatus and tools needed, and compare the sound level of the conventional pump and MHD thruster to show MHD thruster can reduce noise and vibration. Then, the affect of magnetics strength and voltage on the velocity of water produced from the MHD thruster. Last but not least affect of the velocity on the sound level are to be investigate.

There are a number of restrictions that may apply when using a silent MHD thruster for a stealthy micro undersea vehicle. The inclusion of cost elements is thought to go beyond the intended scope of this study, which is one of the main restrictions or scope limitations associated with this study. Only things that are readily available in stores or that have undergone significant prototype testing are allowed to be included in the experiment design.

Approximations were developed based on data for comparable components that may differ in size, capacity, or other features when precise technical or reliability data was not available. The computer model and reliability evaluations take the representative system and component failures into account but do not aim to cover every scenario.

However, an academic analysis of sources of data on certain subjects connected to this study has been investigated. In addition, the literature gives a summary of the present state of knowledge and makes it possible to pinpoint pertinent concepts, procedures, and research requirements to dive deeper into this research.

The magnetohydrodynamic idea is essential for interdisciplinary study, among other things. Pumping materials that are difficult to pump with ordinary pumps is one of this phenomenon's most important applications. This study examines the developments in this field and groups them into various application categories. The most notable MHD application improvements during the preceding 27 years are examined using the literature. In conditions demanding high flow and velocity rates, MHD thrusters may be useful [3].

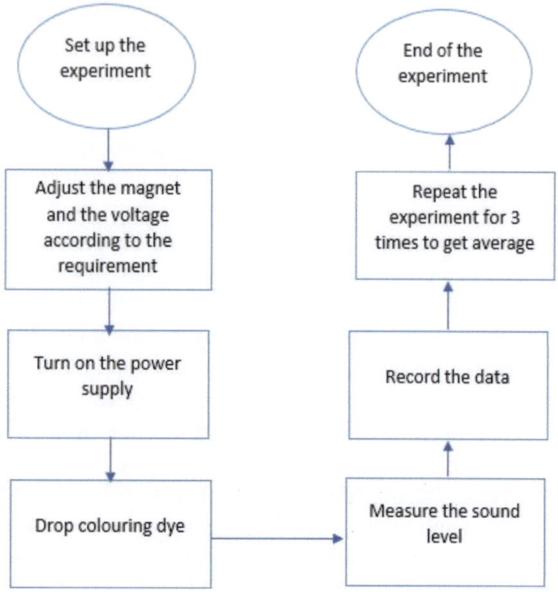

Fig. 16.1 Flow of the experiment comparing conventional and MHD thruster

16.2 Methodology

16.2.1 Flowchart for the Experiment Setup

All the experiments use the same setup but different variables. The flow charts in Figs. 16.1, 16.2, 16.3, and 16.4 describe in brief the flow of every experiment. Some of the experiments use a sound meter, and some do not, depending on the objectives of the experiment. Then, every one of the experiments must be repeated three times to get the average. Also, the coloring dye was dropped to see the flow of water clearly.

16.2.2 Data Recording

Velocity of Water

To calculate the velocity of water, the following equation is used:

$$V = \frac{d}{t}$$

where V is the velocity of the water, d is the distance measured of the water flow in unit centimeters (cm), and t is times taken when the water starts to flow until it

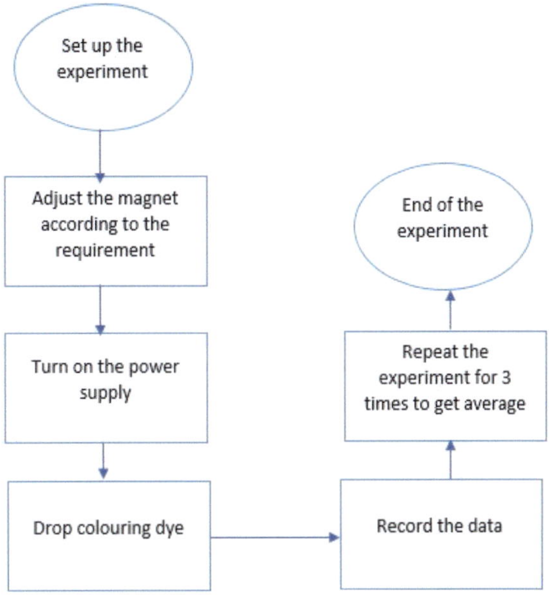

Fig. 16.2 Flow of the experiment of magnetic strength and velocity of water

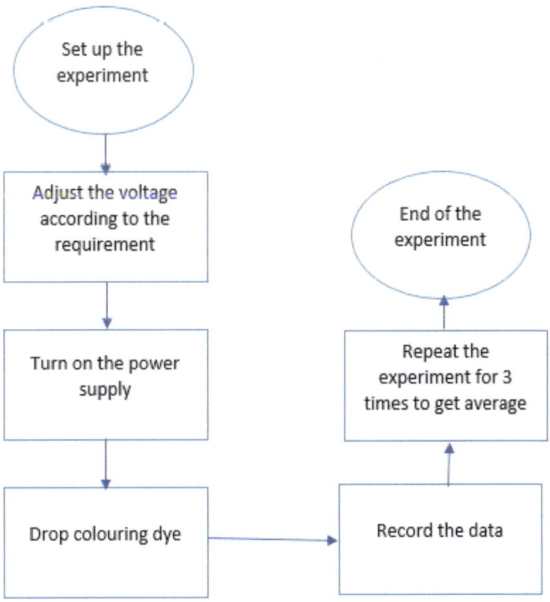

Fig. 16.3 Flow of the experiment of voltage and velocity of water

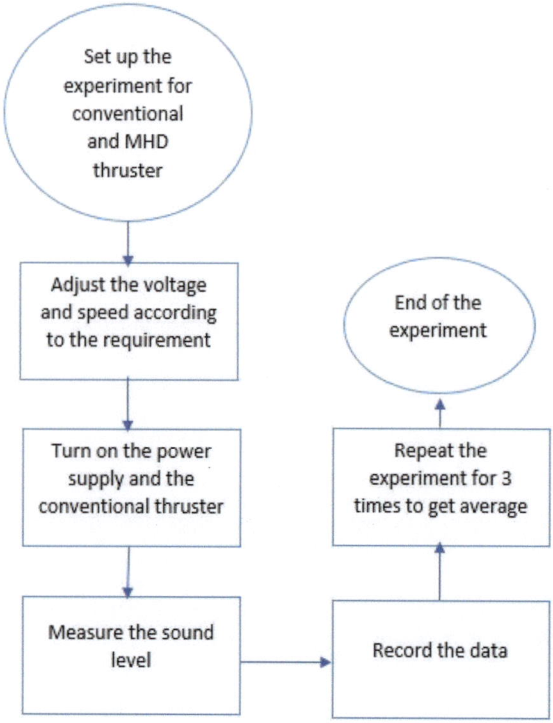

Fig. 16.4 Flow of the experiment of the velocity of water and sound level

reaches the distance in unit seconds (s) that have been measured. The unit for velocity is a meter per second (m/s).

Sound Levels

The data for sound level was collected by measuring the sound using a sound meter. The instrument was placed above the position of the magnet for the MHD thruster experiments, as shown in Fig. 16.5. While the conventional pump was placed above the nozzle where the fluid is coming from. The way to use the instrument is by placing it in the position mentioned above and waiting for 15 s before pressing the max button to get the reading. After that, record the reading. The reading was read three times to get an average reading.

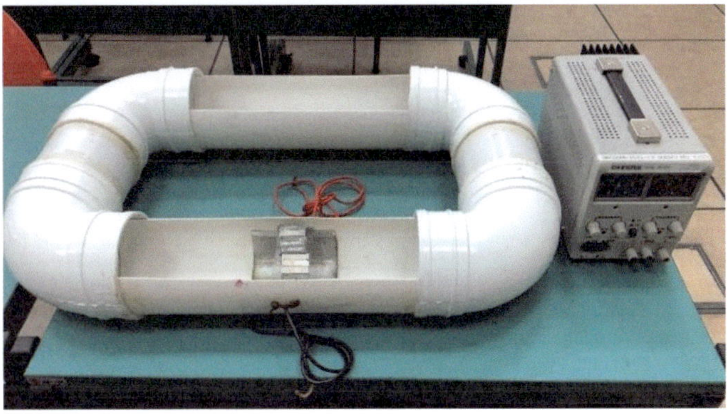

Fig. 16.5 MHD truster

16.3 Results and Discussion

The graphs in Figs. 16.6, 16.7, 16.8, and 16.9 show the results of every experiment. According to the graph, conventional pumps move water faster than MHD thrusters. Conventional pump converts electrical energy to hydraulic energy through mechanical movement. MHD thrusters use magnetic and electric field. In this experiment, sound level is very crucial. The graph shows that MHD thrusters have low sound levels and are quieter than traditional pumps.

The velocity of water moved by the magnetohydrodynamic thruster also depends on the type of water used. For example, this experiment used salt water. Salt water stimulates like seawater. Electrolytic ion exchange can conduct some electricity in seawater [4]. Salt water conducts electricity better [5] than tap water. Salt water is

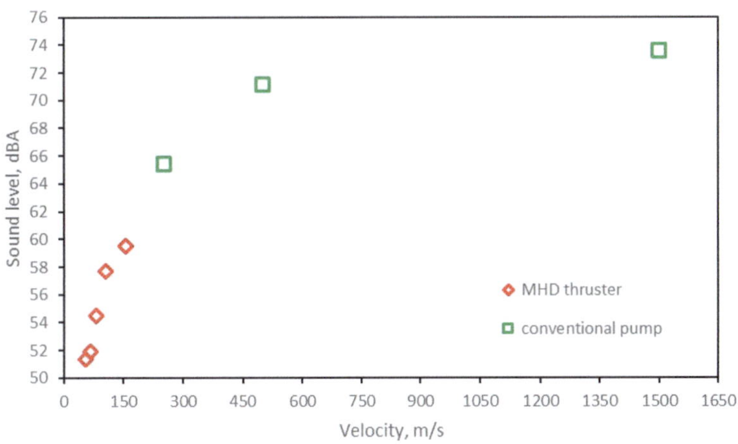

Fig. 16.6 Graphs of comparison of sound level between the conventional pump and MHD thruster

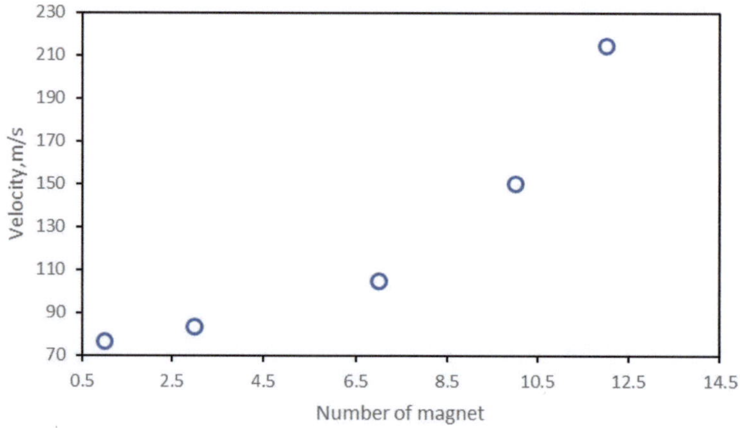

Fig. 16.7 Graphs of numbers of magnet versus velocity of water

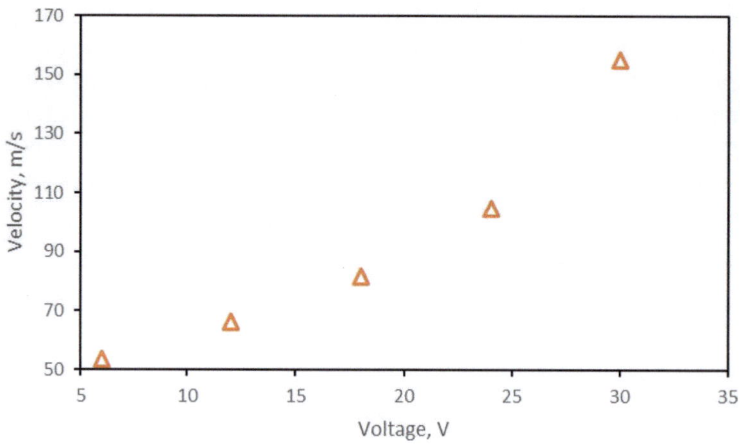

Fig. 16.8 Graphs of voltage versus velocity of water

ionic, while tap water is covalent and poorly conducts electricity. Salt water is denser than tap water.

Magnetohydrodynamic thrusters indeed reduced the noise and vibration compared to conventional pumps. As the magnetic strength and voltage increase, so does the speed of the MHD thrusters [6]. The pattern of the graphs shows that the speed of the MHD thruster is strongly related to the capacity of voltage and magnetic strength [7] being installed. It can be seen clearly in the graph that magnetic strength and voltage increase, and the velocity also increases. As the number of magnets and voltage increase, the velocity of water increases, resulting in a simultaneous increase in sound volume.

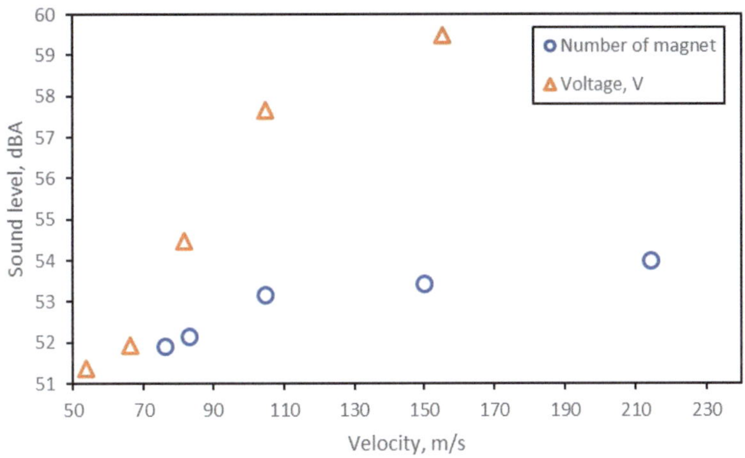

Fig. 16.9 Graphs of water velocity versus sound level

16.4 Conclusions

The study's findings demonstrate that the parameters of the MHD propulsion system can improve water flow. The magnetic field and voltage applied have a significant impact on the thruster's performance. From the experiment has been done and the findings of this study demonstrate that MHD propulsion vessels is marine transport without the need for moving elements as the propulsion system can replaced the mechanical thruster.

The MHD pump resembles a typical water pump in appearance, but it differs in that it has no moving mechanical parts, offers accurate flow control, uses less energy, and produces less dross. These characteristics demonstrate that MHD has a potential feature in the future and can be implemented in nuclear engineering, mining, bioengineering, and maritime propulsion.

References

1. Roy G (1993) Electromagnetic marine propulsion: recent advances and future challenges. In: Proceedings of the third international offshore and polar engineering conference, vol 1, pp 24–27
2. Al-Habahbeh OM, Al-Saqqa M, Safi M, Abo Khater T (2016) Review of magnetohydrodynamic pump applications. Alexandria Eng J 55(2):1344–1358 (Elsevier)
3. Aoki LP, Maunsell MG, Schulz HE (2012) A magnetohydrodynamic study of behavior in an electrolyte fluid using numerical and experimental solutions. Revista de Engenharia Térmica
4. Bui AK, Takeda M, Kiyoshi T (2010) Measurements of flow loss in helical-type seawater MHD power generator. Teion Kogaku (J Cryogenic Soc Jpn) 45(12):506–513
5. Cé Bron D, Viroulet S, Mie Vidal J, Masson J-P, Viroulet P (2017) Experimental and theoretical study of magnetohydrodynamic ship models

6. Chan CS, Cheng JH, Zeng CH, Huang JR, Chen YH, Chen YJ, Pham TT, Chao WH, Jeng JT, Liu TL, Pan KC, Li YH, Chen CY (2020) Design of marine vehicle powered by magnetohydrodynamic thruster. Magnetohydrodynamics 56(1):51–65
7. Ito K, Takahashi T, Fujino T, Ishikawa M (2014) Influences of channel size and operating conditions on fluid behavior in a MHD micro pump for micro total analysis system. J Int Council Electr Eng 4(3):220–226

Chapter 17
Enhancing Vehicle Security: Investigating the Efficacy and Feasibility of Utilizing 7075-T6 Aluminum Alloy as an Anti-intrusion Beam

Adam Harith Halim, Sharmiwati Mohammed Sharif, and Zainal Nazri Mohd Yusuf

Abstract Side impact collisions pose significant risks to passenger safety due to limited deformation space. This study investigates using the 7075-T6 aluminum alloy in vehicle side impact beams to enhance safety. The 7075-T6 alloy's higher specific strength compared to the commonly used 6061 alloy was tested using finite element models and Federal Motor Vehicle Safety Standards. Results showed that the 7075-T6 alloy beam exerted a force three times greater than the 6061 alloy beam, demonstrating superior intrusion resistance and potential for reduced occupant injuries. These findings highlight the promise of 7075-T6 alloy in improving vehicle safety standards and suggest its potential application in future automotive safety designs.

Keywords Side impact · Anti-intrusion beam · 7075-T6 alloy · Crashworthiness

---

A. H. Halim · S. M. Sharif (✉) · Z. N. M. Yusuf
Malaysian Spanish Institute Kulim Hi-Tech Park, Universiti Kuala Lumpur, 09000 Kulim, Kedah, Malaysia
e-mail: sharmiwati@unikl.edu.my

A. H. Halim
e-mail: adam.halim01@s.unikl.edu.my

Z. N. M. Yusuf
e-mail: zainalnazri@unikl.edu.my

S. M. Sharif
Centre for Women Advancement and Leadership, Universiti Kuala Lumpur, 50250 Kuala Lumpur, Wilayah Persekutuan, Malaysia

T. A. A. Razak et al. (eds.), *Progress in Engineering Technology VI*, Advanced Structured Materials 229, https://doi.org/10.1007/978-3-031-81517-1_17

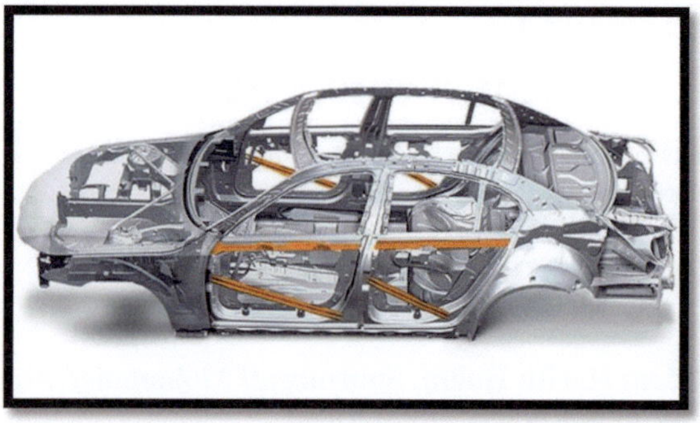

Fig. 17.1 Side impact beams in a car

17.1 Introduction

Crashworthiness is a critical aspect of vehicle safety, emphasizing a vehicle's ability to withstand impacts and protect its occupants. Among various collision scenarios, side impact collisions are particularly perilous due to the minimal deformation space available to absorb the impact energy, which directly affects the safety of passengers [1]. Unlike frontal collisions where bumpers and crumple zones provide substantial protection, side impacts offer limited energy absorption mechanisms, primarily relying on the vehicle's doors [2].

Side impact beams (SIBs) are structural components designed to enhance the side impact resistance of vehicle doors. These beams are essential in absorbing and redistributing the impact energy to minimize intrusion into the passenger compartment. Material selection for SIBs plays a pivotal role in determining their effectiveness [3]. Aluminum alloys, known for their high strength-to-weight ratio, are widely used in the automotive industry for various applications. Among these, the 7075-T6 aluminum alloy, renowned for its superior mechanical properties, emerges as a potential candidate for SIBs, offering an upgrade over the commonly used 6061 alloy [4].

This study aims to explore the efficacy of 7075-T6 aluminum alloy in SIBs by conducting a detailed analysis of its mechanical properties and impact performance through finite element simulations and standardized crash tests. The findings from this study could pave the way for enhanced vehicle safety standards by incorporating high-strength materials in critical structural components (Fig. 17.1).

17.2 Methodology

The methodology of this study involves a comparative analysis of 7075-T6 and 6061 aluminum alloys used in SIBs. This comparison is conducted through finite element analysis (FEA) using the Abaqus software and validated against the Federal Motor Vehicle Safety Standards (FMVSS) 214 protocol [5]. The study focuses on evaluating

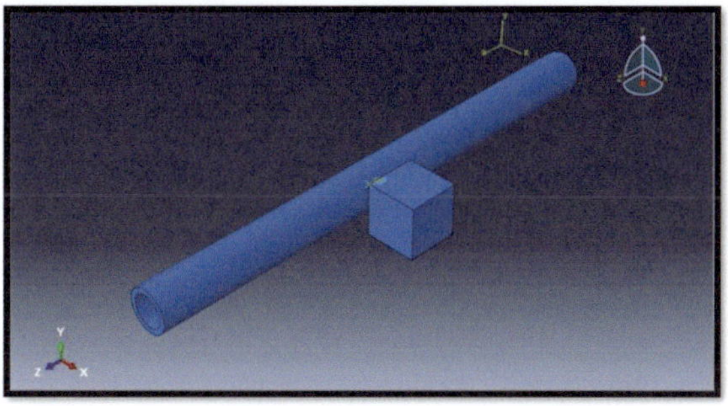

Fig. 17.2 Impact test simulation setup

the impact energy absorption and structural response of different SIB designs under simulated side collision scenarios.

17.2.1 Abaqus Software

Abaqus, a powerful FEA tool, was employed to simulate side impact tests. This software enables precise modeling of material behavior under dynamic loading conditions. The simulation provided detailed insights into displacement, stress distribution, and deformation characteristics of the SIBs, facilitating a comprehensive assessment of their performance.

17.2.2 Impact Test

Impact tests were conducted following the FMVSS 214 protocol, which specifies the requirements for side impact protection in vehicles. The tests involved a solid square block impacter striking the SIB at a speed of 50 km/h. This setup replicates real-world collision dynamics, allowing for an accurate evaluation of SIB deformation and energy absorption capabilities (Fig. 17.2).

17.2.3 Boundary Condition

In the FEA model, the ends of the SIBs were immobilized to replicate the welded connection points in a car door. The impacter moved at a velocity of − 50 km/h in the X-direction, simulating the impact force experienced during a side collision. This boundary condition ensured a realistic simulation of the SIB's response to impact (Fig. 17.3).

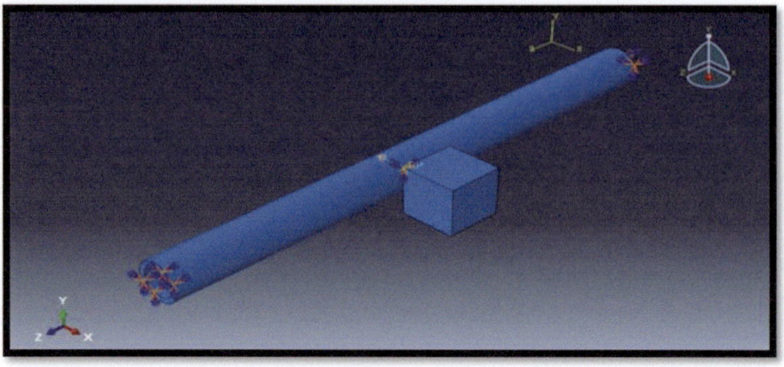

Fig. 17.3 Boundary Condition for the simulation

17.2.4 Meshing

Meshing is a critical step in FEA, influencing the accuracy and reliability of the simulation results. A coarse mesh with uniform quadrilateral elements was utilized to balance computational efficiency and result accuracy. The meshing properties are detailed in Table 17.1.

17.3 Results and Discussion

The results of the deformation and displacement analysis revealed that the 7075-T6 alloy beams exhibited significantly higher impact resistance compared to the 6061 alloy beams. Figure 17.4 visually compares the deformation patterns of the 6061 alloy (left) and 7075-T6 alloy (right) beams under impact conditions. The images show that the 7075-T6 alloy beam undergoes significantly less deformation compared to the 6061 alloy beam, further illustrating the superior impact resistance of the 7075-T6 alloy. The reduced deformation in the 7075-T6 alloy beam is consistent across all stages of impact, highlighting its strength and ability to withstand higher forces without compromising structural integrity. This visual evidence supports the quantitative results showing that the 7075-T6 alloy can endure a maximum force of

Table 17.1 An overview of the meshing properties employed in the analysis

Element size	13 mm
Size function	Uniform
Methods	Quadrilaterals
Number of nodes	7728
Number of elements	5651

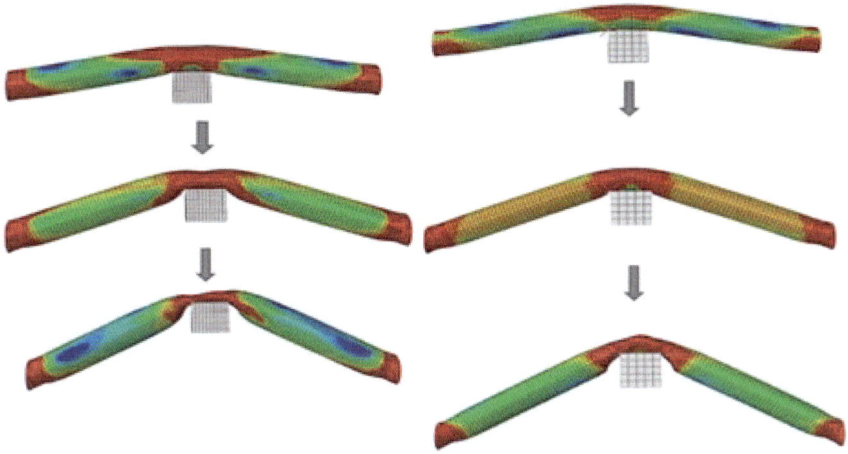

Fig. 17.4 Comparison between 6061 alloy (left) and 7075-T6 alloy (right)

3.55×10^5 N, significantly higher than the 6061 alloy's maximum force of 1.16×10^5 N.

The force–displacement graph (Fig. 17.5) further illustrates the enhanced performance of the 7075-T6 beam. The graph shows that the 7075-T6 beam endured higher forces and exhibited greater energy absorption, emphasizing its potential to improve SIB performance and occupant safety in side collisions.

17.3.1 Modified SIB Structural Deformation Comparison

Three different cross-sectional designs (rectangle, circular, and square) were tested to evaluate their deformation behavior and impact performance (see Fig. 17.6). Each design offered unique benefits in terms of stress distribution and impact resistance.

17.3.2 Design A

The deformation patterns of the 6061 beam (left) and 7075-T6 beam (right) under impact conditions in Fig. 17.7 clearly illustrate the superior structural integrity of the 7075-T6 alloy. The 7075-T6 beam demonstrates significantly less plastic deformation compared to the 6061 beam, indicating its higher resistance to stress and its ability to better maintain its shape under impact.

The square tube beams showed significant plastic deformation under impact. The 7075-T6 beam exhibited higher stress resistance (6.47×10^8 Pa) compared to the 6061 beam (2.2×10^8 Pa). The force–displacement graph (Fig. 17.8) indicated

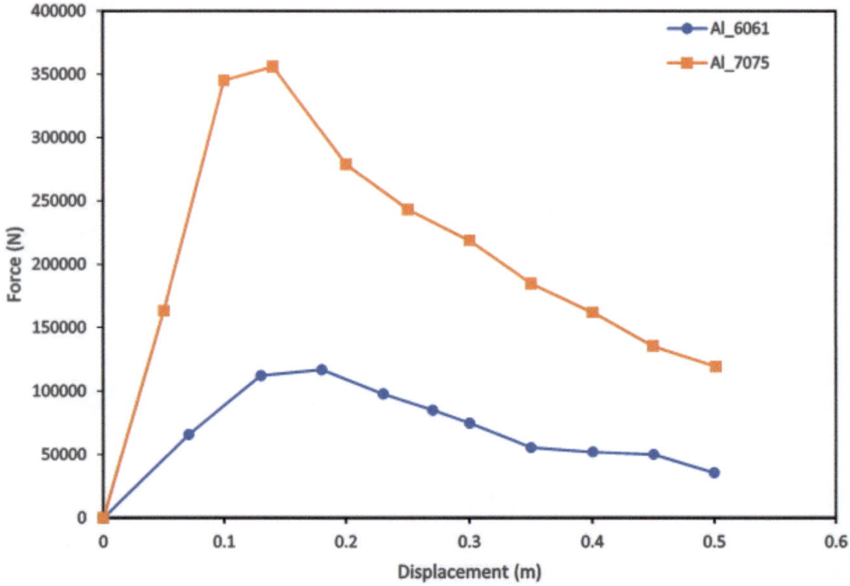

Fig. 17.5 Force displacement graph comparison between 7075-T6 and 6061 beam

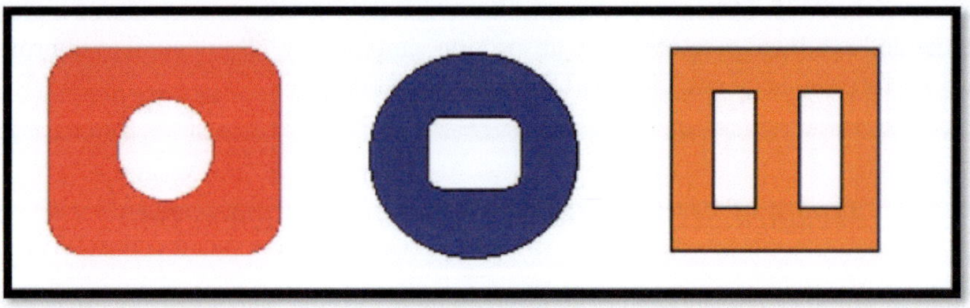

Fig. 17.6 The modified beam (from left: design A, design B, and design C)

Fig. 17.7 Deformation of 6061 beam (left) and 7075-T6 beam (right)

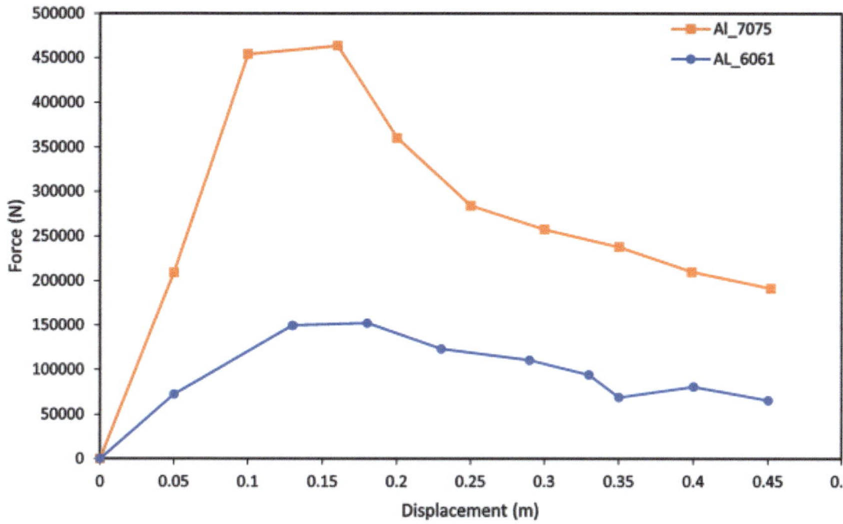

Fig. 17.8 Force–displacement graph comparison of design A

that the 7075-T6 beam endured nearly three times the force of the 6061 beam, demonstrating superior strength and impact energy absorption.

17.3.3 Design B

Figure 17.9 shows a comparison between two circular beams, likely from a structural or mechanical simulation. The beams are color-coded, indicating stress distribution, with red areas representing higher stress.

The graph in Fig. 17.10 illustrates the force–displacement relationship for two aluminum beams, 7075 and 6061. The 7075 beam exhibits significantly higher force

Fig. 17.9 Deformation of 6061 beam (left) and 7075-T6 beam (right)

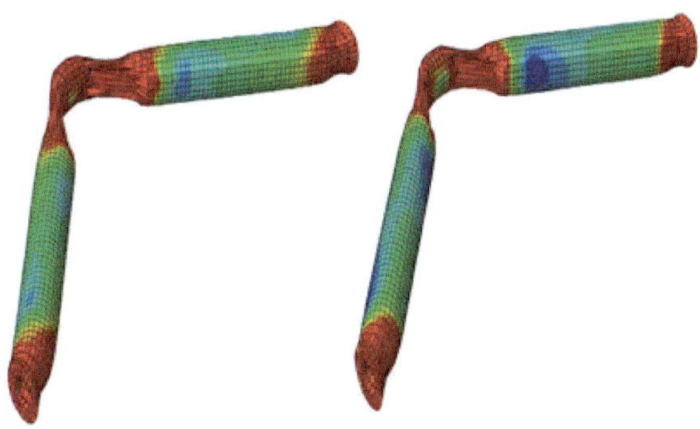

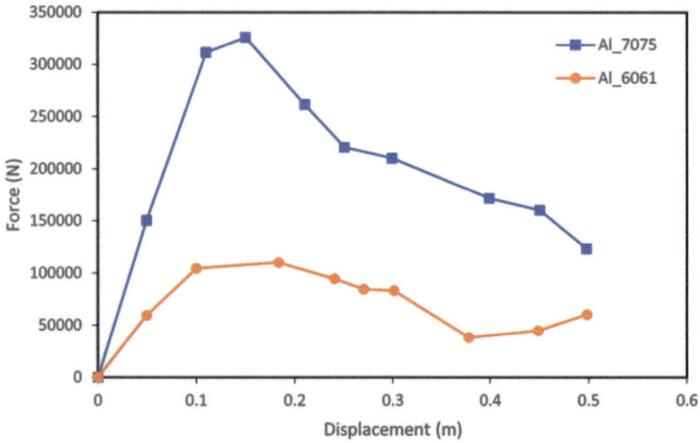

Fig. 17.10 Force–displacement graph comparison of design B

resistance, peaking at approximately 326,000 N at 0.15 m displacement, compared to the 6061 beam, which peaks around 100,000 N. This indicates that the 7075 beam can withstand greater forces before substantial deformation occurs, offering better impact performance. After reaching peak force, both beams show a decline, with the 7075 beam displaying a more gradual reduction, further emphasizing its superior structural integrity under continued loading and reduced intrusion, which enhances its overall impact performance.

17.3.4 Design C

Figure 17.11 shows two simulation results comparing the deformation and stress distribution of beams under load. The color coding (ranging from blue to red) represents different levels of stress, with red indicating areas of high-stress concentration. The simulation results indicate that 6061 beam experiences more significant deformation and stress concentration than the other, suggesting a lower resistance to impact or load. This visual comparison aligns with previous observations that the 7075-T6 beam demonstrates superior performance in resisting deformation and distributing stress evenly, which enhances the structural integrity and impact performance of the beam.

The force–displacement graph in Fig. 17.12 demonstrates that the 7075-T6 beam exhibits superior force resistance, reaching a peak force of approximately 3.26×10^5 N, significantly higher than the 6061 beam. The 7075-T6 beam also maintains a more gradual force reduction with increasing displacement, indicating enhanced stress distribution and reduced deformation. This combination of strength and flexibility makes the 7075-T6 beam an optimal choice for SIB applications.

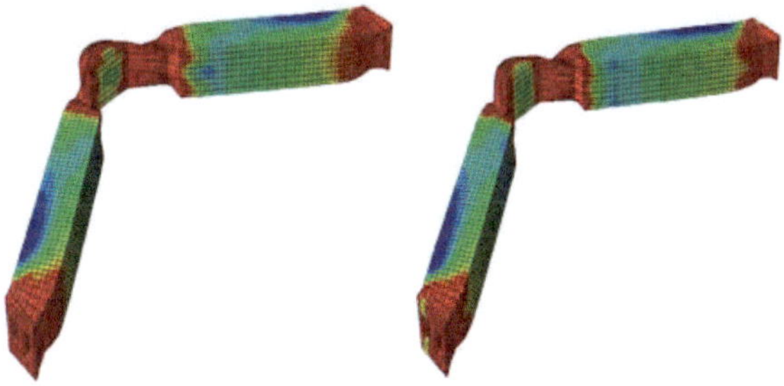

Fig. 17.11 Deformation of 6061 beam (left) and 7075-T6 beam (right)

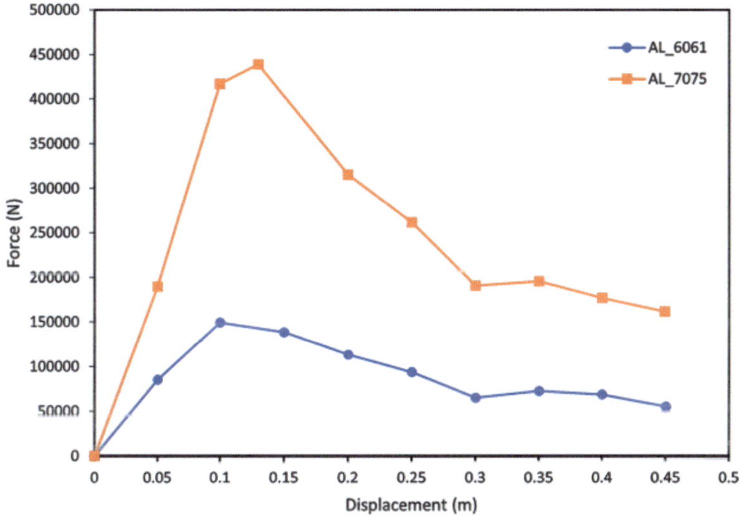

Fig. 17.12 Force–displacement graph comparison of design C

The enhanced performance of the 7075-T6 alloy can be attributed to its higher specific strength and superior mechanical properties compared to the 6061 alloy. The results from the FEA simulations and impact tests indicate that the 7075-T6 alloy not only provides better resistance to deformation but also improves the energy absorption capacity of the SIBs. The different cross-sectional designs further illustrate that the geometric configuration of the beams plays a crucial role in determining their impact performance. The square and circular designs, in particular, showed significant promise in uniformly distributing stress and withstanding higher forces.

17.4 Conclusion

This study demonstrated that the 7075-T6 aluminum alloy significantly outperforms the 6061 alloy in SIB applications. The 7075-T6 alloy's superior strength and impact resistance make it a promising material for enhancing vehicle crashworthiness and occupant safety. Modified SIB designs, especially the 'I' beam and circular with square tube configurations, showed optimal performance in withstanding impact forces. These findings underscore the importance of material selection and innovative design in advancing vehicular safety standards. The use of 7075-T6 aluminum alloy in side impact beams could lead to substantial improvements in vehicle safety, potentially reducing the severity of injuries in side collisions and saving lives.

Acknowledgements This research was supported by Universiti Kuala Lumpur. We thank our colleagues from Universiti Kuala Lumpur Malaysian Spanish Institute who provided insight and expertise that greatly assisted the research.

References

1. Dong G, Wang D, Zhang J, Huang S (2007) Side structure sensitivity to passenger car crashworthiness during pole side impact. Tsinghua Sci Technol 12:290–295
2. Lin J, Bao X, Hou Y et al (2020) Investigation on yield behavior of 7075–T6 aluminum alloy at elevated temperatures. Chin J Mech Eng 33:76
3. Rebelo PM (2019) Design study of a side intrusion beam for automotive safety. Master's thesis. Instituto Superior Técnico, pp 1–10
4. Du Z, Deng Z, Cui X, Xiao A (2021) Deformation behavior and properties of 7075 aluminum alloy under electromagnetic hot forming. Materials (Basel) 14(17):4954
5. National Highway Traffic Safety Administration (NHTSA) (2012) Laboratory test procedure for FMVSS No. 214, dynamic side impact protection moving deformable barrier test requirements using advanced test dummies. Report No. 45
6. Veeraswamy K, Venkata Sudheer Babu V (2016) Design and analysis of a composite beam for side impact protection of a car door. Int Res J Eng Technol 2(1):1–6
7. Fildes BN, Lane JC, Lenard J, Vulcan AP (1994) Passenger cars and occupant injury: Side impact crashes. Accident Research Centre, Monash University, pp 1–76
8. Abdollah MF, bin Hassan R (2013) Preliminary design of side door impact beam for passenger cars using aluminium alloy. J Mech Eng Technol 5:11–18
9. Ben G, Aoki Y, Sugimoto N (2007) Impact properties of CFRP/Al hybrid beam for absorbing impact energy in side collision of automobiles. In ICCM international conference on composite materials

Chapter 18
Eco-Composite Using an Eggshells Reinforced Epoxy Matrix for Automotive Parts

Abdul Jawad Mohd Said, Sharmiwati Mohammed Sharif, Noorhelinahani Abu Bakar, Ana Salwa Ramli, and Nurrasyidah Izzati Rohaizat

Abstract This project focuses on the investigation of composite materials employing an epoxy matrix combined with waste materials from natural sources, specifically utilizing eggshells as an environmentally friendly composite component. The specimens were categorized into three groups: A, B, and C, each with distinct compositions. Category A encompassed specimens consisting of 20% eco-composite and 80% epoxy matrix. In category B, epoxy content comprised 75% while the eco-composite constituted 25%, and in category C, epoxy constituted 70% while the eco-composite formed 30% of the composition. Additionally, the project utilized sieves to control particle size, with sizes of 1.03 mm, 0.8 mm, and 0.651 mm. Upon analysis, the findings indicate that among the examined specimens, those containing 70% epoxy and 30% eco-composite, featuring a particle size of 0.651 mm, exhibited superior tensile strength and hardness. These promising mechanical attributes hold significance for applications in designing automotive components. This study's insights into the mechanical properties and structural characteristics of the composite material hold potential for informing the development of car parts, enhancing both performance and sustainability in the automotive industry.

A. J. M. Said
Malaysian Spanish Institute Kulim Hi-Tech Park, Universiti Kuala Lumpur, 09000 Kulim, Kedah, Malaysia
e-mail: jawad.mohd@s.unikl.edu.my

S. M. Sharif (✉) · N. A. Bakar · A. S. Ramli · N. I. Rohaizat
Centre for Women Advancement and Leadership, Universiti Kuala Lumpur, 50250 Kuala Lumpur, Wilayah Persekutuan, Malaysia
e-mail: sharmiwati@unikl.edu.my

N. A. Bakar
e-mail: noorhelinahani@unikl.edu.my

A. S. Ramli
e-mail: anasalwa@unikl.edu.my

N. I. Rohaizat
e-mail: nurrasyidah.izzati@unikl.edu.my

T. A. A. Razak et al. (eds.), *Progress in Engineering Technology VI*, Advanced Structured Materials 229, https://doi.org/10.1007/978-3-031-81517-1_18

Keywords Material · Composite · Eggshells · Car spoiler · Car fender

18.1 Introduction

This study is centered around the investigation of composite materials utilizing an epoxy matrix in conjunction with waste materials derived from natural sources [1]. Specifically, eggshells were employed as an eco-friendly composite element within this research initiative. The specimens under scrutiny were categorized into three distinct groups: A, B, and C, each characterized by unique compositional ratios. Category A encompassed specimens consisting of a composite-to-epoxy ratio of 20:80. In category B, the epoxy content constituted 75%, juxtaposed with an eco-composite content of 25%. Similarly, category C featured specimens with an epoxy composition of 70%, coupled with an eco-composite content of 30%. Furthermore, particle size management was achieved through the utilization of sieves with aperture dimensions of 1.03 mm, 0.8 cmm, and 0.651 cmm.

18.2 Methodology

This section details the methodologies and procedures used to develop an eco-composite with an eggshell-reinforced epoxy matrix for automotive parts [2]. The process included designing and fabricating a tensile test mold in SolidWorks, generating a 3D-printed mold to ASTM 368 specifications, and producing a silicone mold.

Specimens were composed of eggshells, epoxy resin, and hardener, creating nine distinct samples using the Taguchi method in Minitab software. The samples were divided by particle size using three sieves (0.651, 0.80, and 1.03 mm) and varied eggshell content (20, 25, and 30%). Specimens 1A, 2A, and 3A contained 2 g of eggshells, 1B, 2B, and 3B had 2.5 g, and 1C, 2C, and 3C contained 3 g, as summarized in Table 18.1.

The composite mixtures were cured at room temperature (20–25 °C) for 24 h. Figures 18.1, 18.2, and 18.3 show the cured specimens. After curing, the specimens were removed from the molds and subjected to tensile testing using a universal testing machine (UTM) in accordance with ASTM 638 guidelines.

Mechanical assessments, including ultimate tensile strength, modulus of elasticity, and elongation, were conducted on the eggshell composite specimens at varying speeds to comprehensively evaluate their mechanical properties. This methodology facilitates the evaluation of the mechanical performance of the resulting composite materials for automotive applications.

Table 18.1 Specimens with particle sizes

No.	Parameter		Particle size
	Epoxy (%)	Eco-fiber (%)	
1A	80	20	0.651
1B	75	25	0.651
1C	70	30	0.651
2A	80	20	0.800
2B	75	25	0.800
2C	70	30	0.800
3A	80	20	1.030
3B	75	25	1.030
3C	70	30	1.030

Fig. 18.1 Specimen with particle size 0.651 mm

18.3 Results and Discussions

Analyzing microstructural properties is essential for understanding the mechanical and functional behavior of materials. In this study, advanced microscopy with a 9.740 mm magnification lens was used to examine the composites formed by mixing eggshells and epoxy. This detailed examination revealed significant variations in microstructure across different specimens, highlighting the complex interplay between composition, processing conditions, and resulting material characteristics. These findings emphasize the need for comprehensive microstructural analysis to fully understand composite performance. The discussion will explore these

Fig. 18.2 Specimen with particle size 0.80 mm

Fig. 18.3 Specimen with particle size 1.03 mm

microstructural differences and their implications for the mechanical and functional properties of composites (Figs. 18.4, 18.5, 18.6, 18.7, 18.8, 18.9, 18.10, 18.11 and 18.12).

The results of the experimental data were meticulously analyzed to elucidate the intricate relationship between tensile strength and hardness. The investigation sought to uncover the influence of varying particle concentrations on the mechanical properties of the specimens. Notably, the findings underscore a compelling trend: an augmented particle concentration correlates positively with enhanced tensile strength. This phenomenon was particularly pronounced in the context of the present

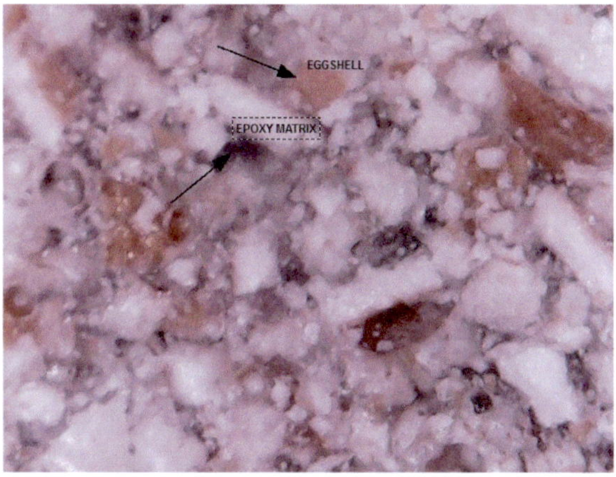

Fig. 18.4 Microstructure of specimen 1A

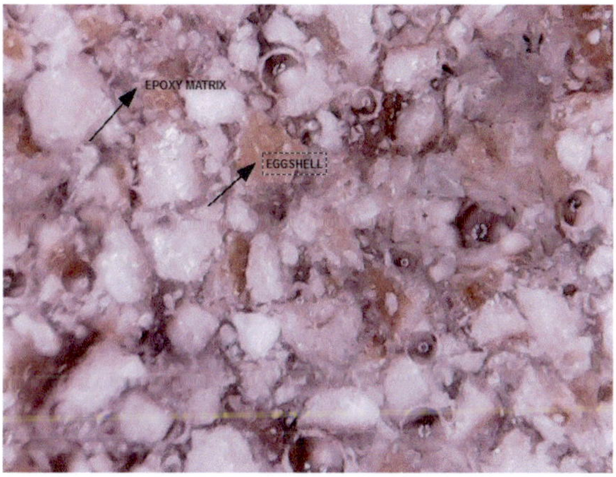

Fig. 18.5 Microstructure of specimen 1B

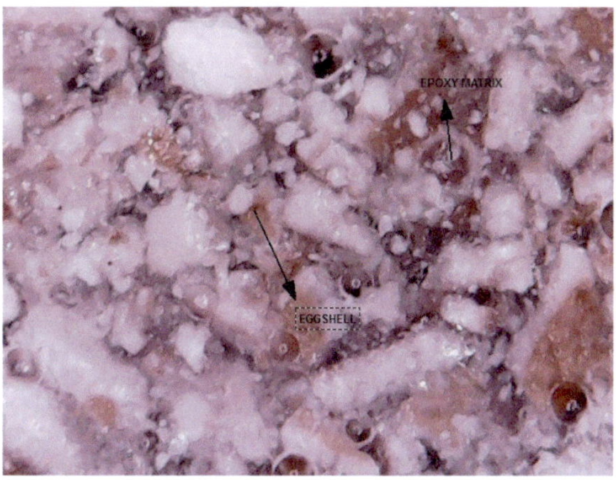

Fig. 18.6 Microstructure of specimen 1C

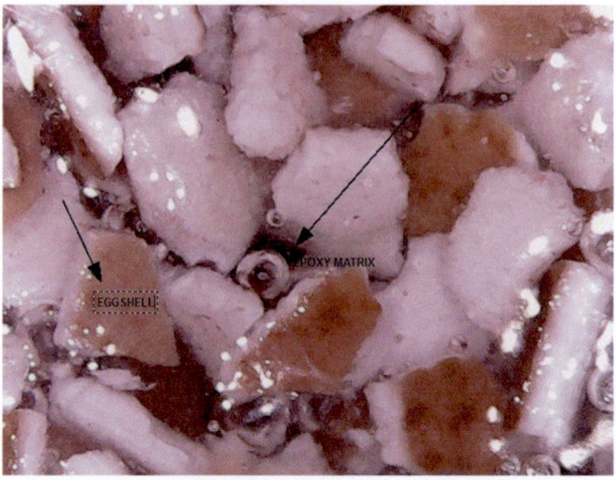

Fig. 18.7 Microstructure of specimen 2A

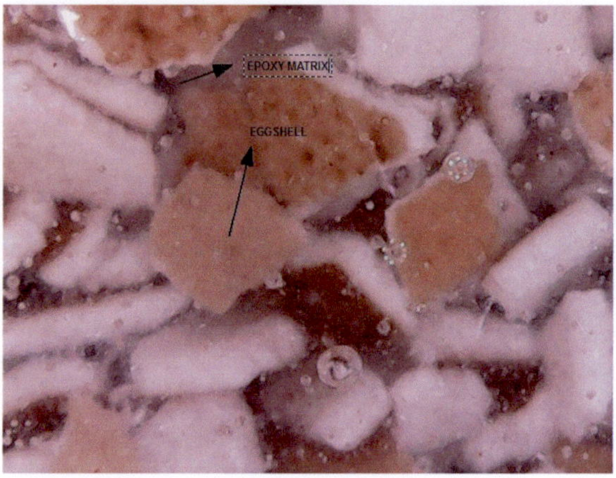

Fig. 18.8 Microstructure of specimen 2B

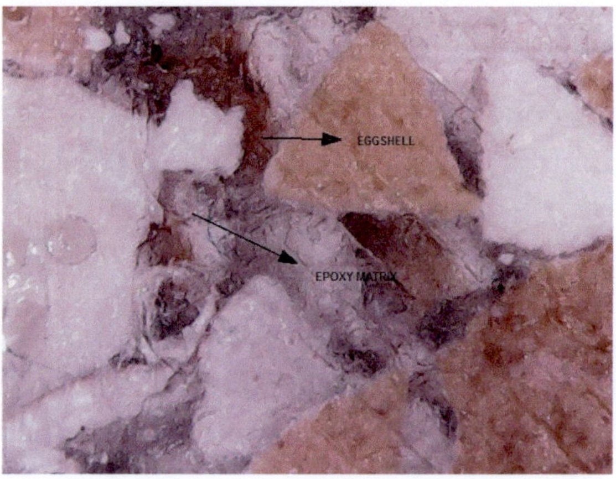

Fig. 18.9 Microstructure of specimen 2C

Fig. 18.10 Microstructure on specimen 3A

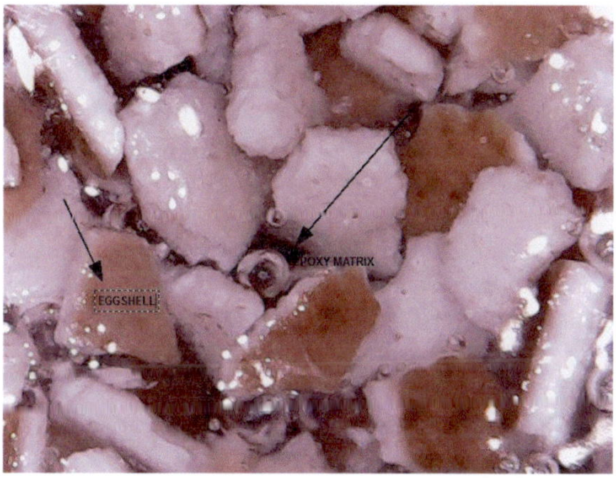

Fig. 18.11 Microstructure on specimen 3B

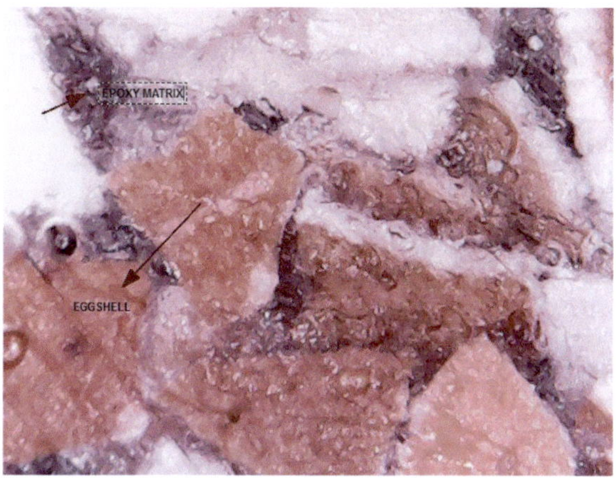

Fig. 18.12 Microstructure on specimen 3C

study, where specimens incorporating 30% particle content exhibited a remarkable increase in tensile strength. Intriguingly, the results further elucidate the impact of particle size on mechanical performance [3]. Among the specimens subjected to the same 30% particle concentration, those with a smaller particle size of 0.651 mm exhibited the highest tensile strength, followed closely by those featuring a particle size of 1.03 mm. This observation lends credence to the notion that smaller particle sizes, when combined with a higher particle concentration, engender greater specimen robustness. The confluence of these outcomes not only substantiates the central hypothesis but also offers insights into the intricate interplay between material composition and mechanical performance. In essence, the results accentuate the potential for tailoring material properties through judicious manipulation of particle size and concentration, thereby opening new avenues for advanced material design and engineering applications [2] (Table 18.2).

The experimental analysis depicted in Fig. 18.13 shows the relationship between particle percentage and maximum tensile strength. The highest tensile strength of 16.35 N/mm^2 was achieved with 30% particle content and a particle size of 0.651 mm, highlighting the significant influence of both factors. Additionally, as particle percentage increased, so did the hardness of the specimens. This trend suggests that particle incorporation enhances mechanical properties, likely due to improved interlocking and reinforcement within the material structure (Fig. 18.14).

The observed relationship between particle application and material properties shows potential for tailoring material characteristics through controlled particle incorporation. These findings are significant for engineering applications where enhanced tensile strength and hardness are desirable. Further investigations are needed to optimize particle size and applied percentage for a wider range of materials.

This study developed a novel material using insights from tensile tests. The simulation illustrated its application in automotive components, specifically a car

Table 18.2 Result of tensile test

No.	Parameter		Particle size	Tensile (N/mm^2)	Hardness (N/mm^2)	Young's modulus (N/mm^2)
	Epoxy (%)	Eco-fiber (%)				
1A	80	20	0.651	12.15	3.521	918.49
1B	75	25	0.651	12.79	3.71	504.1
1C	70	30	0.651	16.35	4.74	1210.82
2A	80	20	0.80	7.35	2.13	654.93
2B	75	25	0.80	8.32	2.41	1597.85
2C	70	30	0.80	8.69	2.52	1372.14
3A	80	20	1.03	12.92	3.75	2135.64
3B	75	25	1.03	14.9	4.32	2891.71
3C	70	30	1.03	15.64	4.53	2120.37

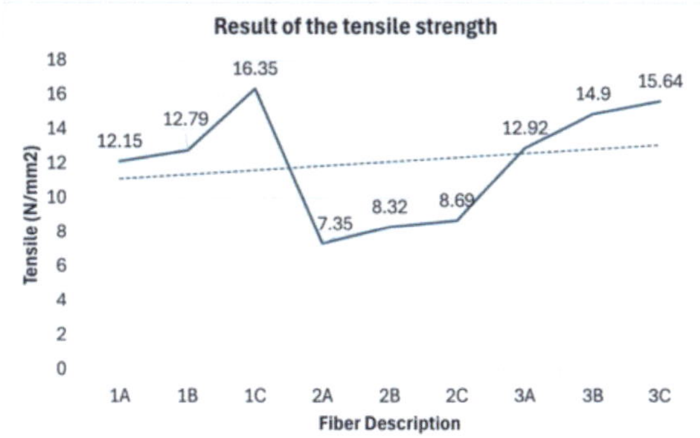

Fig. 18.13 Tensile strength of the nine specimens that have undergone testing

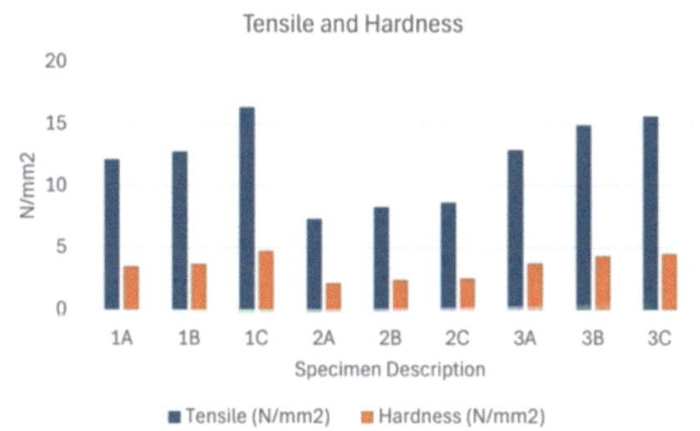

Fig. 18.14 Tensile versus Brinell hardness

spoiler, subjected to a static load of 2000 N. The innovative eggshell-derived material showed enhanced mechanical properties. A comparative evaluation between the eggshell material and conventional material revealed notable differences in structural behavior, suggesting the eggshell material's potential for automotive applications.

The collected data detail the simulation outcomes, focusing on the performance disparity between the original spoiler material and the eggshell material. The study also examined the impact of incorporating eggshell material onto the car fender, using stress and displacement values as key performance indicators [3].

The simulation results showed that ABS material had a stress of 2.167 MPa and a displacement of 3.749 mm, while the eggshell material had the same stress but a higher displacement of 6.192 mm as shown in Fig. 18.15. For the car fender, stress and displacement were 3.619 MPa and 2.663 mm, respectively, indicating reasonable flexibility for both components as shown in Fig. 18.16.

However, a collision could lead to complete fracturing of the spoiler and fender. Strain analysis showed 7.87% for ABS and a much lower 1.30% for the eggshell material as shown in Fig. 18.17, indicating greater resilience for the latter. These findings suggest that eggshell material is promising for certain automotive components, though further research is needed to explore its full potential.

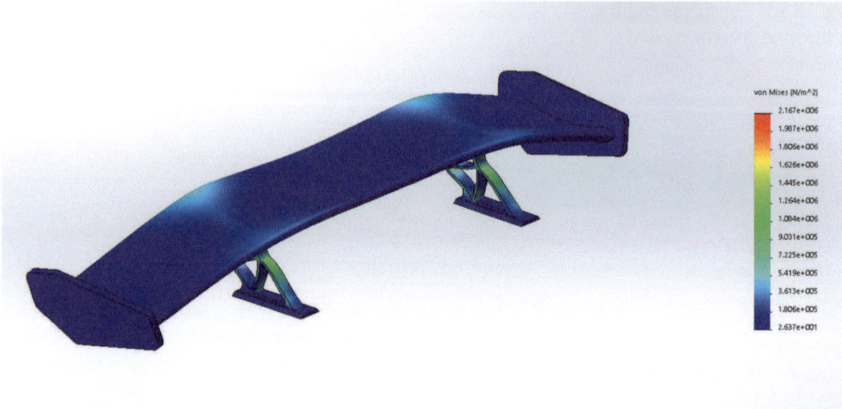

Fig. 18.15 Stress result on ABS material

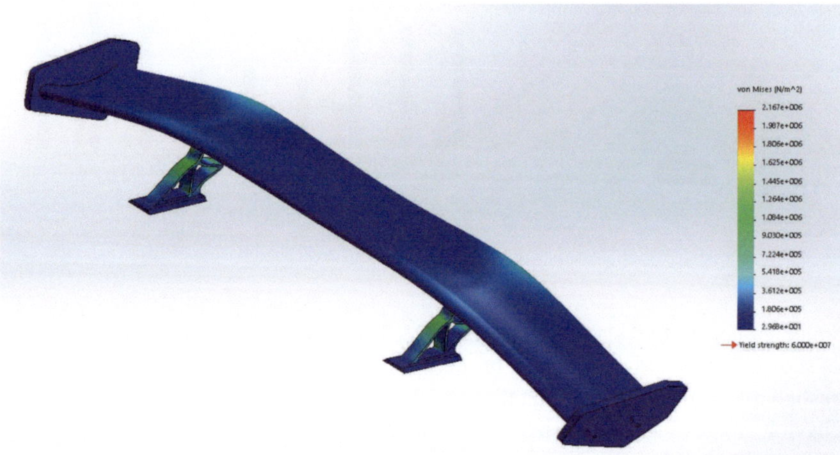

Fig. 18.16 Stress result on eggshell material

The research findings reveal that among the various examined specimens, those composed of 70% epoxy and 30% eco-composite, possessing a particle size of 0.651mm, demonstrated exceptional tensile strength and hardness characteristics. These pivotal mechanical attributes hold profound implications for the prospective design and fabrication of automotive components. The study's comprehensive insights into the mechanical properties and structural nuances of the composite material hold substantial promise for guiding advancements in the realm of car part engineering, fostering an amalgamation of enhanced performance and sustainable practices within the automotive sector.

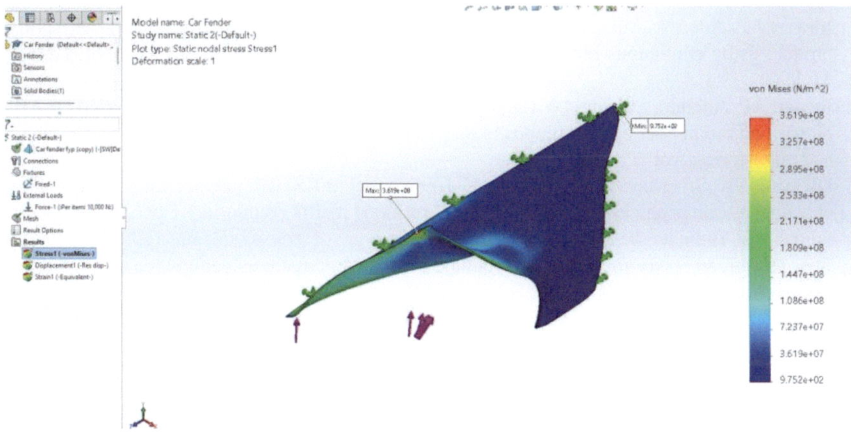

Fig. 18.17 Stress result on car fender

18.4 Conclusion

Through meticulous investigation into variations in fiber weight distribution, the study has successfully shed light on the intricate relationship between composite composition and tensile properties. The outcomes clearly demonstrate that epoxy compounds enhanced with this composite formulation exhibit superior tensile strengths in comparison to those reinforced with alternate composite materials, irrespective of their size. The pivotal role of the weight fraction in influencing the tensile properties has been substantiated, highlighting the paramount configuration of epoxy at 70% and fiber at 30%, characterized by a particle size of 0.651 mm, which attains the peak tensile strength. Despite these promising findings, it is evident that further refinement is imperative before contemplating the integration of eggshell-based composites within the automotive industry. This study serves as a significant stepping stone toward the realization of innovative materials that could revolutionize the field, emphasizing the pressing need for continued research and development in this domain.

Acknowledgements This research was supported by Universiti Kuala Lumpur. We thank our colleagues from Universiti Kuala Lumpur Malaysian Spanish Institute who provided insight and expertise that greatly assisted the research.

References

1. Bogoeva-Gaceva G, Avella M, Malinconico M, Buzarovska A, Grozdanov A, Gentile G, Errico ME (2007) Natural fiber eco-composites. Polym Compos 28(1):98–107

2. Irawan AP, Anon A, Sukania IW, Undefined U (2020) Manufacturing process of car spoiler product using continuous rattan fiber composite materials. Int J Mech Eng Technol (IJMET) 11(1)
3. Irawan AP, Adianto, Sukania IW (2020) Tensile strength of car spoiler product based on ABS plastic and rattan fiber epoxy composite materials. In: IOP conference series: materials science and engineering, vol 725, no 1
4. Dwiwedi SK, Srivastava AK, Sugimoto K, Chopkar M (2018) Microstructural and mechanical characterization of chicken eggshell-reinforced Al6061 matrix composites. Open J Metal 08(01)
5. Saidah A, Yudinata Y, Susilowati SE (2018) Design of composite material of rice straw fiber reinforced epoxy for automotive bumper. In: 3rd International conference on computing, engineering, and design, ICCED 2017 (2018-March)
6. Petrasek S, Muller M (2017) Polymeric particle composites based on filler from hen eggshells. Eng Rural Dev 16:175–180

Chapter 19
Development of an Autonomous Floating Vehicle Model with a Flood Detection System

Siti Rohana Binti Ahmad, Ainatul Fatimah Binti Abdul Mukti, Mohamad Sazali Bin Said, and Muhammad Zunnurrin Bin Ghazali

Abstract The autonomous vehicle with a flood detection system is a crucial advancement in disaster management and environmental monitoring. It uses sensors to measure water levels and meteorological parameters, enabling safe navigation in turbulent floodwaters. The system also provides data for effective long-term flood management strategies. This innovative approach could save lives, reduce property damage, and improve flood resilience, transforming disaster preparedness and response in flood-prone regions.

Keywords Flood · Water lever rising · Autonomous vehicle · Floating mechanism · Environmental monitoring

S. R. B. Ahmad (✉) · A. F. B. A. Mukti
Mechanical Section, Malaysian Spanish Institute Kulim Hi-Tech Park, Universiti Kuala Lumpur, 09000 Kulim, Kedah, Malaysia
e-mail: sitirohana@unikl.edu.my

Centre for Women Advancement and Leadership, Universiti Kuala Lumpur, 50250 Kuala Lumpur, Wilayah Persekutuan, Malaysia

A. F. B. A. Mukti
e-mail: ainatul.mukti19@s.unikl.edu.my

M. S. B. Said
Manufacturing Section, Malaysian Spanish Institute Kulim Hi-Tech Park, Universiti Kuala Lumpur, 09000 Kulim, Kedah, Malaysia
e-mail: msazali@unikl.edu.my

M. Z. B. Ghazali
R&Z Auto, 470 Jalan Kota Kenari 3, 09000 Kulim, Kedah, Malaysia

T. A. A. Razak et al. (eds.), *Progress in Engineering Technology VI*, Advanced Structured Materials 229, https://doi.org/10.1007/978-3-031-81517-1_19

19.1 Introduction

Floods are devastating natural disasters exacerbated by climate change, urbanization, and deforestation. To mitigate their impact and respond swiftly, innovative flood monitoring and early warning systems are needed. The development of an autonomous floating vehicle with a flood detection system is a significant step forward. Traditional methods, relying on manual measurements, often fail to provide timely data, leaving communities vulnerable to flooding [1].

Autonomous vehicles are revolutionizing environmental monitoring by collecting real-time data across various domains. In flood monitoring, an autonomous floating vehicle with a flood detection system aims to create a versatile platform capable of navigating through floodwaters in turbulent conditions. The vehicle features sensors for water levels, rainfall intensity, temperature, humidity, and other meteorological parameters, along with a robust navigation system [2].

The project aims to create an efficient flood monitoring system using an autonomous vehicle. The vehicle collects real-time data on rising water levels and precipitation rates, predicting flood events and generating timely warnings. It uses advanced data processing techniques and energy-efficient design, combining battery power with solar panels for continuous operation.

The autonomous vehicle, combining robotics, hydrology, meteorology, and data science, is a crucial tool for flood monitoring. Its interdisciplinary approach could save lives, reduce property damage, and improve disaster management strategies in flood-prone areas [3].

Therefore, the development of an autonomous floating vehicle with a flood detection system model is a visionary endeavor that holds the promise of transforming our approach to flood monitoring and disaster response. This technology has the potential to enhance the safety and resilience of communities vulnerable to flooding and represents a vital step forward in addressing the growing challenges posed by climate change and extreme weather events.

19.2 Literature Review

19.2.1 Autonomous Vehicle

Autonomous vehicles, also known as self-driving cars, utilize advanced technologies like sensors, cameras, radar, Lidar, GPS, and software algorithms to operate and navigate autonomously. Autonomous vehicles are increasingly popular, enabling researchers to optimize their studies by incorporating autonomous systems, which are intelligent machines capable of performing tasks independently, eliminating human control, and overcoming limitations [4]

Self-driving cars are progressively gaining market penetration. While there were about 31 million machines with some level of automation in operation around the world in 2019, that number is predicted to rise to 54 million by 2024. From that, the usage of autonomous vehicle systems is believed that it can benefit more people or users onwards [3, 5, 6].

In this study, this autonomous vehicle system will be helpful in solving the problem of drivers affected by the rise of water in their areas. It also can reduce the panic of the driver when sudden disaster or chaos occurs because this autonomous system will operate without any human effort [7].

19.2.2 Sensor Overview in Amphibious Vehicle (AV)

Light Detection and Ranging Sensor

LiDAR is a distant sensing technique that works on the principle of producing infrared or laser light pulses that reflect from target objects. The equipment detects these reflections, and the time between emission and reception of the light pulse allows for distance estimate. LiDAR sensors produce data in the form of a series of points, also known as point cloud data (PCD), in 1D, 2D, and 3D areas, as well as object intensity information. The PCD comprises the x, y, and z coordinates as well as the intensity information of the obstacles inside the scene or surrounds for 3D LiDAR sensors [6].

Control System and Microcontroller

Microcontrollers, in general, contain processor cores, memory, and programmable input/output peripherals. They are used in automatically controlled products and devices, such as automobile engine control systems, remote controls, office machines, appliances, and other embedded systems [8]. Arduino is open source with low-cost hardware. While the Arduino microcontroller comes in unique features depending on the complexity of use, Arduino mega is used for this work because it allows us to use more pins and more memory to retain codes [9, 10].

19.2.3 Floating Mechanism

During flood or rainy seasons, safety airbags and a floating buoy are used to prevent drivers from sinking into water. The buoy's mechanism is similar to safety airbags, with water level detected as the actuator. The buoy automatically opens based on the situation, ensuring the vehicle remains afloat on the water surface [11].

19.2.4 Water-Level Detection

This project uses a water level sensor system based on the Internet of Things (IoT), allowing objects to transfer data over the internet without human interaction [12]. The system design includes hardware components like sensors, actuators, microcontrollers, and communication devices, while a mobile application like Blynk IoT will connect with the sensor and provide user notifications. IoT is strongly associated with machine-to-machine communication [13].

19.2.5 Uses of Radar

To monitor the river's water speed, water level, and other hydrological information in real time, and reduce the loss of people's lives and property caused by flood disasters, radar can also be used to monitor the water level. Dual radars are used to measure water level and flow separately. This system has the advantages of flexibility, easy deployment, and fast response and can be widely used in flood prevention and emergency rescue and other occasions [14].

19.3 Methodology

19.3.1 Development of Autonomous Vehicle Model

The mini car model is created to replace the real car because it is difficult, and the scale is too big to test the rough concept of this research on the actual car, which is shown in Fig. 19.1. The sedan car type will be used in this research.

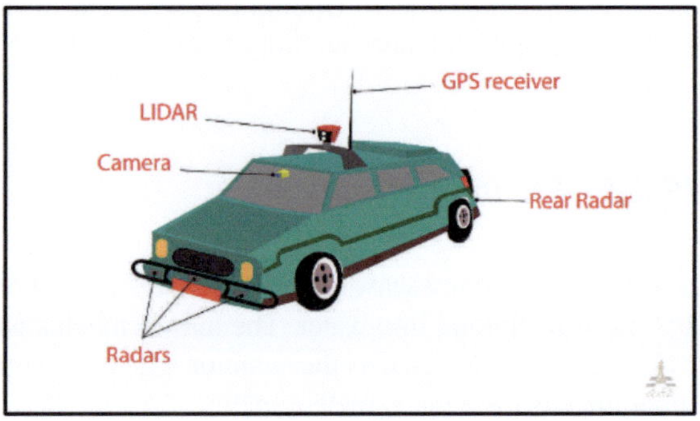

Fig. 19.1 Car model concept

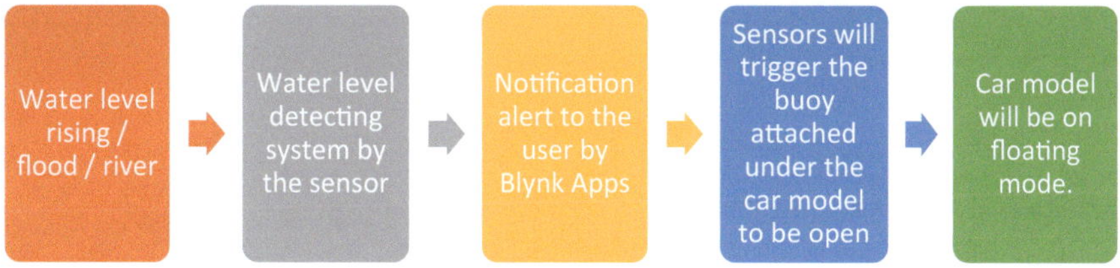

Fig. 19.2 Flowchart of how the floating mechanism works

19.3.2 Floating Mechanism

Figure 19.2 shows the flowchart of how the floating mechanism works in this project.

19.3.3 Mechanical Part

After the floating mechanism is activated, the autonomous system will next take over to move the floating car to a safer place. Sensors will be the pilot to control the movement of the car. The car model will use the propeller as the drone concept when in floating mode/water. Mini fan blades will be attached at 4 sides of the car model (like the tire). The car will choose its direction according to the sensor's detection.

19.4 Result and Discussion

19.4.1 Design of the AV Model Car

All the drawings were carried out using AutoCAD. Figure 19.3 shows the 2D and 3D drawing. The 2D drawing (orthographic) indicates 3 views from top, side, and front of the model of autonomous vehicles.

General specification of the autonomous vehicle model is that the length is 25 cm (about 9.84 in.), width is 17 cm (excluding tire and gap), and height is 8 cm (about 3.15 in.). After the design is finalized, the next stage is to develop the model of the autonomous vehicle.

19.4.2 Developed Model

Figure 19.4 shows the autonomous vehicle that is equipped with flood detection. It indicates the top view of the model, the usage of Arduino as microcontroller of the

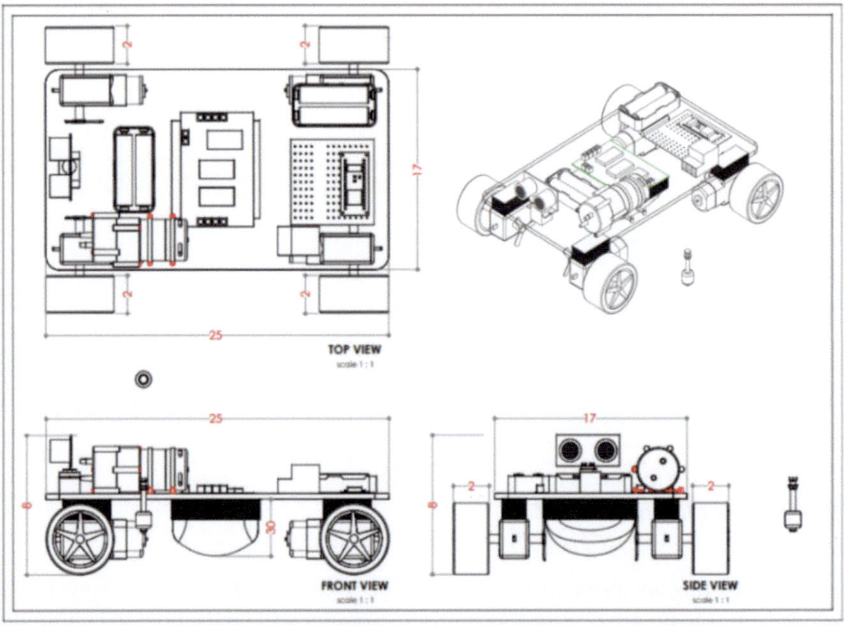

Fig. 19.3 2D and 3D drawings of the model

model to control the autonomous movement of the car and another microcontroller board to execute the water level sensor and the pump that inflated the buoy. From the front view of the car model, propeller attached to the tire can be seen on both front and rear tires. The buoy attached to the model can be seen at the bottom of the model, which was inflated by the pump triggered by the float sensor when water rising or being detected.

19.4.3 Result

Table 19.1 shows the data collected and observed from the overall testing of the model developed. The testing has been set up five times in a water-effect test rig. The time taken for the buoy to be activated when the water rise was recorded to analyze the functionality of the floating mechanism installed. Next, the time taken for the model to complete the testing with constant distance was recorded to analyze its velocity. The velocity is analyzed in its distribution to show the stability of the model functioning.

From Fig. 19.5, the lower line graph shows the time taken for the buoy to activate when the water level has been detected by the sensor on the model. Overall, the time taken for the floating mechanism to be activated is around 50–65 s which can be said

Fig. 19.4 The Autonomous vehicle model

Table 19.1 Data recorded from testing process

Test	Floating Mechanism—Time taken to activate (s)	Time taken to complete the testing (s)	Distance (m)	Velocity (m/s)
1	63	175	0.72	3.77
2	56	168	0.79	3.93
3	49	149	1.00	4.43
4	58	171	0.76	3.86
5	51	155	0.92	4.26
Average	56.5	165.75	0.82	4.00
Standard deviation	5.59	11.08	0.12	0.28

to be under 1 min. It proves that the system functions well as the limitation from this study is that the system will be delayed around 30 s as the component used is a mini and low-cost version.

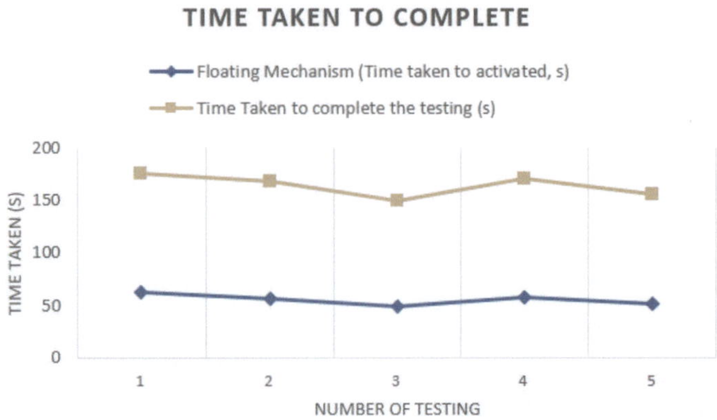

Fig. 19.5 Time taken to complete testing

Next, the upper line graph shows the time taken for the model to complete the testing with a constant distance test rig of 0.66 m. The time taken recorded is about 150 to 180 s (about 3 min) range which is under 3 min of functionality. The time recorded simplifies the stability of the model developed as it shows a steady linear progress for the 5-testing data, and the difference from each data is about 4–26 s only which is under 30 s.

19.5 Conclusions

This project aimed to develop an autonomous vehicle with a flood sensor system mini model to reduce engine maintenance costs during rainy seasons or floods. The model was tested to ensure its application to actual car models with specific parameters. The design process involved researching components, selecting suitable ones, and analyzing their functions using simulation. The model was completed with a mini car chassis and a floating mechanism or flood sensor system. However, the components were not waterproof and needed to be insulated manually. The model was tested in water-effect conditions, with the floating mechanism tested for functionality and stability. The project's concept has proven successful and can be implemented in real-life applications to provide aid during floods or rainy seasons.

References

1. Meckel D, Keen H, Heupel C, Berns K (2022) Transferring off-road control concepts to watercraft used in flooded areas. Presented at the

2. Liu B, Ayaz F, Anh HD, Edwards O, Zeng Y, Sheng Z, Duan X, Tian D (2023) UAV integrated Internet-of-Vehicle (IoV) system for flooding monitoring and rescue. In: Lecture notes in electrical engineering
3. Ignatious HA, El-Sayed H, Khan M, Mokhtar B (2022) Realizing situation awareness in Autonomous. SSRN Electron J. https://doi.org/10.2139/ssrn.4128843
4. Frejek M, Nokleby S (2008) Design of a small-scale autonomous amphibious vehicle. In: Canadian conference on electrical and computer engineering (2008)
5. Ignatious HA, El-Sayed H, Khan MA, Mokhtar BM (2023) Analyzing factors influencing situation awareness in autonomous vehicles—A survey
6. Ignatious HA, El Sayed H, Khan M (2021) An overview of sensors in Autonomous vehicles. Procedia Comput Sci (2021)
7. Ignatious HA, El-Sayed H, Khan MA (2022) Sensor technology for autonomous vehicles. In: Encyclopedia of sensors and biosensors, vol 1–4, 1st edn.
8. Dang L, Sriramoju N, Tewolde G, Kwon J, Zhang X (2017) Designing a cost-effective Autonomous Vehicle Control System Kit (AVCS Kit). In: 2017 IEEE AFRICON: science, technology and innovation for Africa, AFRICON 2017
9. Asafa TB, Afonja TM, Olaniyan EA, Alade HO (2018) Development of a vacuum cleaner robot. Alexandria Eng J 57. https://doi.org/10.1016/j.aej.2018.07.005
10. Annuar KAM, Dahari RM, Harun MH, Sapiee MRM, Ab Hadi NA (2018) Design and implementation of vacuum cleaner robot using Arduino and smartphone. J Eng Appl Sci 13. https://doi.org/10.3923/jeasci.2018.5692.5698
11. Odetti A, Altosole M, Caccia M, Viviani M, Bruzzone G (2018) Wetlands monitoring: hints for innovative autonomous surface vehicles design. In: Technology and science for the ships of the future—Proceedings of NAV 2018: 19th international conference on ship and maritime research
12. Younan M, Houssein EH, Elhoseny M, Ali AA (2020) Challenges and recommended technologies for the industrial internet of things: a comprehensive review. Measurement (Lond) 151. https://doi.org/10.1016/j.measurement.2019.107198
13. Perwej Y, Ahmed M, Kerim B, Ali H (2019) An Extended review on Internet of Things (IoT) and its promising applications. Commun Appl Electron 7. https://doi.org/10.5120/cae2019652812
14. Wang T, Hao S, Ma X (2021) Water velocity and level monitoring based on UAV radar. In: IMCEC 2021—IEEE 4th advanced information management, communicates, electronic and automation control conference

Chapter 20
Experimental Study of a Lab-Scale Extractor of Pyrolysis Oil from Waste Tyres

Siti Rohana Binti Ahmad, Nazhan Affandy Bin Aznan, Mohamad Sazali Bin Said, and Muhammad Zunnurrin Bin Ghazali

Abstract The improper disposal of waste tyres is a significant environmental issue, and pyrolysis, a thermal decomposition process, offers a promising solution for sustainable management. This project focuses on designing and developing a lab-scale extractor to produce pyrolysis oil from waste tyres. After 2D and 3D drawings were created using SolidWorks, the extractor was set up and tested. At an optimum power of 2000 W, the mass of pyrolysis oil obtained was about 70% higher than for 1000 W. However, the tyre residues decreased by 45%, indicating that more pyrolysis oil can be produced with a small amount of waste. This extractor addresses the need for a compact, efficient, and cost-effective system to explore the feasibility of tyre pyrolysis oil.

Keywords Lab-scale extractor · Pyrolysis · Waste tyre · Pyrolysis oil

---

S. R. B. Ahmad (✉)
Mechanical Section, Universiti Kuala Lumpur, Malaysian Spanish Institute Kulim Hi-Tech Park, 09090 Kulim, Kedah, Malaysia
e-mail: sitirohana@unikl.edu.my

S. R. B. Ahmad · N. A. B. Aznan
Centre for Women Advancement and Leadership, Universiti Kuala Lumpur, 50250 Kuala Lumpur, Wilayah Persekutuan, Malaysia
e-mail: nazhan.aznan16@s.unikl.edu.my

M. S. B. Said
Manufacturing Section, Universiti Kuala Lumpur, Malaysian Spanish Institute Kulim Hi-Tech Park, 09090 Kulim, Kedah, Malaysia
e-mail: msazali@unikl.edu.my

M. Z. B. Ghazali
R&Z Auto, 470 Jalan Kota Kenari 3, 09000 Kulim, Kedah, Malaysia

T. A. A. Razak et al. (eds.), *Progress in Engineering Technology VI*, Advanced Structured Materials 229, https://doi.org/10.1007/978-3-031-81517-1_20

20.1 Introduction

Waste tyre disposal has emerged as a global environmental challenge, driven by the sheer volume of discarded tyres and the associated ecological and health risks [1]. Traditional methods of tyre disposal, such as landfilling and incineration, not only contribute to environmental pollution but also squander valuable resources [2]. To mitigate these issues and harness the untapped potential of waste tyres, there is a growing need for innovative and sustainable solutions. One such solution is pyrolysis, a thermochemical process that converts waste tyres into valuable pyrolysis oil, among other valuable byproducts [3, 4].

Pyrolysis involves the thermal decomposition of organic materials in the absence of oxygen, resulting in the generation of pyrolysis oil, char, and gas [5]. Pyrolysis oil is a versatile resource that can be used as a feedstock for various industrial applications, including energy generation, chemicals production, and transportation fuels [6, 7]. However, the successful implementation of pyrolysis as a tyre recycling method relies on robust research and development efforts, which are often hindered by the lack of accessible lab-scale extractors for pyrolysis oil production [8, 9].

This project addresses this critical gap by introducing a lab-scale extractor specifically designed to produce pyrolysis oil from waste tyres. The lab-scale extractor serves as a vital tool for researchers and scientists to explore the feasibility of tyre pyrolysis, investigate process parameters of the resulting pyrolysis oil. By providing a compact, efficient, and cost-effective system, this extractor facilitates in-depth studies into tyre pyrolysis processes and contributes to the broader goal of sustainable waste tyre management.

In this project, a lab-scale extractor of pyrolisis oil was designed and developed. The project's focus is to replicate industrial-scale pyrolysis processes while maintaining the flexibility and accessibility needed for laboratory research. The pyrolisis oil and its residue mass were measured at different power levels between 1000 and 2000 W. To embark on this journey of tyre pyrolysis research at the laboratory scale, this project aims to not only advance the understanding of waste tyre recycling but also contribute to sustainable practices that reduce environmental harm and promote the efficient use of valuable resources. This lab-scale extractor will empower researchers and scientists to explore the promising realm of pyrolysis as a solution to one of the most pressing environmental challenges of our time: the responsible and sustainable management of waste tyres.

20.2 Methodology

The project started with three concept designs (sketching) followed by a morphological chart to identify the best tool and equipment for the fabrication process. Next, the best sketching design was selected for 2D and 3D drawings using SolidWorks. The specimen used in this experiment was tyre powder which was obtained from

R&Z Auto and Tyre, Kulim, Kedah. The tyre was obtained from old waste tyres. The furnace is made up of a stainless-steel pressure cooker with 110 mm diameter, 300 mm depth and half inch thickness of ceramic fibre insulation. The source of heat comes from a 2000 W induction cooker. The piping has been made by using copper rod. The condenser is made of stainless steel. Stainless steel is easily cooled down when cold water is supplied to the inlet and circulates through the outlet thus enhancing the condensation of vapour produced from the furnace to obtain pyrolysis oil.

20.2.1 Testing Procedure

The experiment involved weighing 250 g of powdered tyre and feeding it into a furnace with a 1000 W induction stove. The experiment was conducted for 30 min, and the gases were channelled to a water-cooled condenser. The condensate and residue were collected, and the oil was measured. The experiment was repeated three times, using different power values (1250, 1500, 1750, and 2000 W) at the induction cooker. The average reading was obtained.

20.3 Result and Discussion

Figure 20.1 shows the selected design of the lab-scale extractor for pyrolysis oil from waste tyres in 2D drawings. Figure 20.2 shows the 3D drawings of the design using SolidWorks. The actual lab setup is shown in Fig. 20.3.

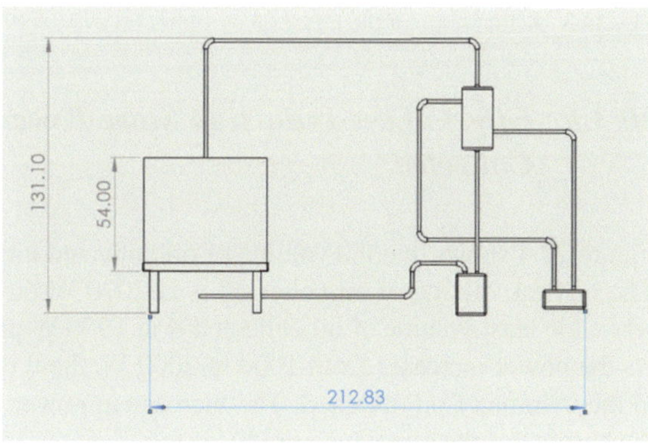

Fig. 20.1 The 2D drawing of the lab-scale pyrolysis oil extractor

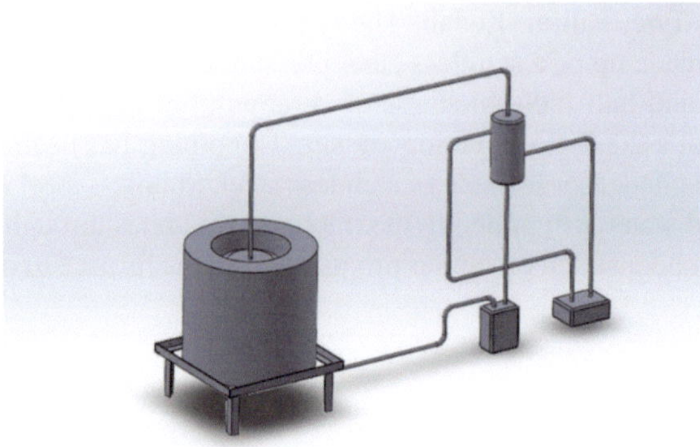

Fig. 20.2 The 3D layout of the extractor

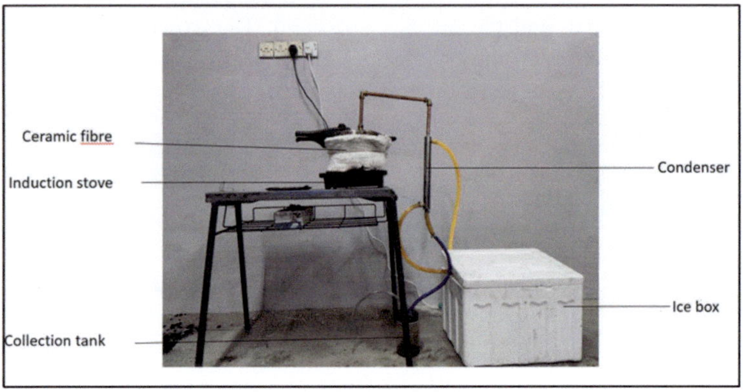

Fig. 20.3 Actual setup of the lab-scale pyrolysis oil extractor

20.3.1 Effect of the Induction Stove Power on Volume of Oil Collected

Figure 20.4 shows that the volume of oil collected increases as the power increases. The highest volume of oil collected is at 2000 W power setting which is 12.47 ml while the least volume of oil collected is at 1000 W power setting which is 3.73 ml. As the power increases from 1000 to 2000 W, there is an increase of about 70.01% in the volume of oil obtained. The increase in power also increases the heat rate of the burning of the tyre which produces more vapour to be condensed thus producing more oil.

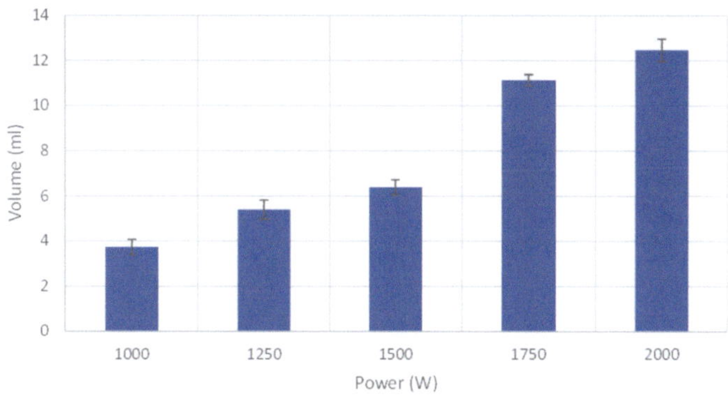

Fig. 20.4 Volume of oil collected versus power

20.3.2 Effect of Induction Stove Power on Mass of Oil Collected

The pattern of this graph in Fig. 20.5 is the same as the pattern of volume of oil obtained versus power which shows that the higher the power of the induction stove the higher the mass of the oil collected. The highest value of the mass of oil obtained is at 2000 W power setting which is 19.03 g while the lowest is at 1000 W power setting which is 5.7 g. The percentage difference of the mass of oil as the power increase from 1000 to 2000 W is about 70.05%. The pattern and the percentage difference are almost the same because the mass is obtained by weighing the oil.

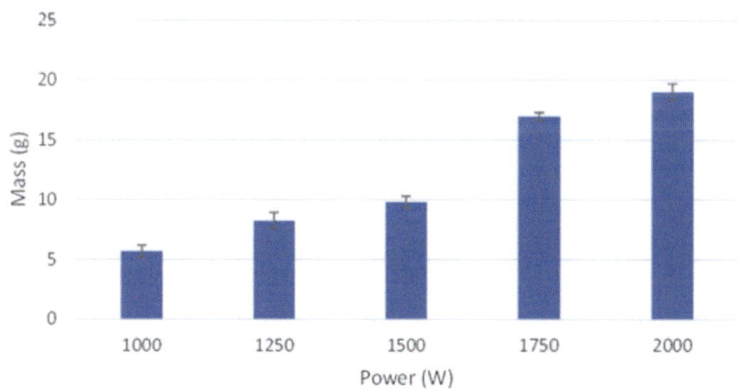

Fig. 20.5 Mass of oil collected versus power

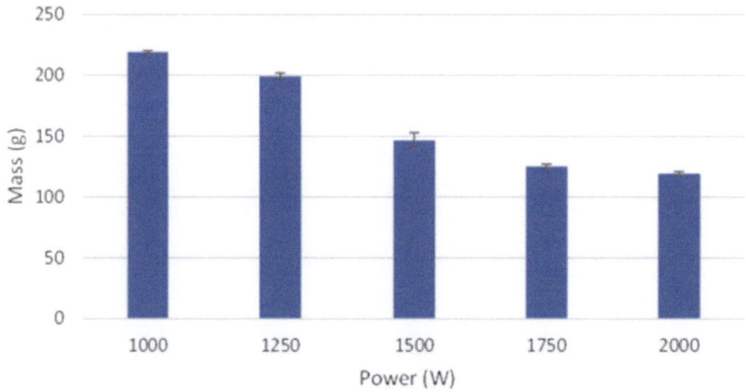

Fig. 20.6 Mass of residue versus power

20.3.3 The Effect of the Induction Stove Power on the Mass of Residue

Figure 20.6 shows the decreasing trend which means that the higher the power of the induction stove, the lower the mass of residue. The highest value of mass of residue is 218.97 g at 1000 W power. While the lowest value is 119.47 g at 2000 W power. As the power increases from 1000 to 2000 W, the mass of residue will decrease about 45.44%. It can be said that higher power value increases the weight loss rate of tyre powder which means the pyrolysis process occurs faster.

20.3.4 Percentage of the Mass of Oil and the Mass of the Residue

Figure 20.7 indicates that as power increases the percentage of mass of oil obtained increases while the percentage of mass of residue decreases. The highest percentage of mass of residue is at 1000 W which is 87.59% while the lowest is at 2000 W power setting which is 47.78%. The highest percentage of mass of oil collected is at 2000 W power setting which is 7.61% while lowest is at 1000 W power setting which is 2.28%. This is because the pyrolysis process really relies on the heat rate at the reaction chamber. The higher the heat rate, the faster the pyrolysis process which increases the percentage of products obtained while reducing the percentage of residue.

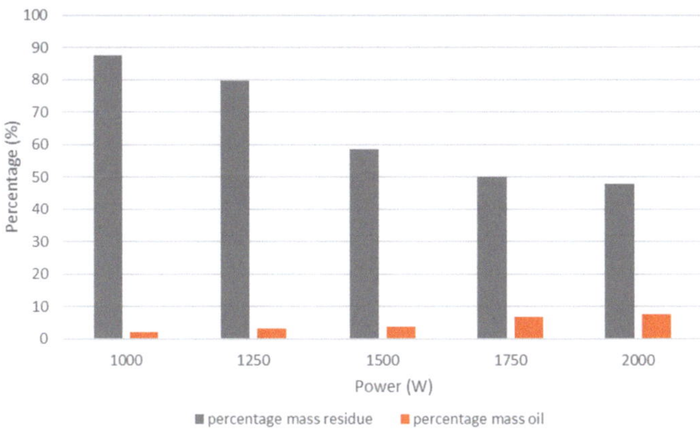

Fig. 20.7 Comparison between the percentage of mass of oil obtained and mass of residue

20.4 Conclusions

The lab-scale pyrolysis oil extractor layout was designed using 2D and 3D SolidWorks. The setup was carefully chosen and the effect of power changes on the percentage of pyrolysis oil and residue produced was analyzed. As the power increased from 1000 to 2000 W, the volume and mass of oil increased by about 70%. The higher the power of the induction stove, the lower the mass of residue tyre powder by up to 45%. This lab-scale pyrolysis oil extractor can efficiently produce pyrolysis oil from waste tyres using 2000 W, assisting in the disposal of old worn tyres and turning them into valuable products. Future research aims to explore the potential of this extractor to generate high-quality pyrolysis oil samples, enabling researchers to assess its composition, energy content, and impurities, paving the way for innovative applications.

References

1. Zhang G, Chen F, Zhang Y, Zhao L, Chen J, Cao L, Gao J, Xu C (2021) Properties and utilization of waste tire pyrolysis oil: a mini review
2. Sharuddin SDA, Abnisa F, Daud WMAW, Aroua MK (2018) Pyrolysis of plastic waste for liquid fuel production as prospective energy resource. In: IOP conference series: materials science and engineering
3. Czajczyńska D, Krzyżyńska R, Jouhara H, Spencer N (2017) Use of pyrolytic gas from waste tire as a fuel: a review. Energy 134. https://doi.org/10.1016/j.energy.2017.05.042
4. Hoang AT, Nguyen TH, Nguyen HP (2020) Scrap tire pyrolysis as a potential strategy for waste management pathway: a review
5. Aldayyat EA, Saidan MN, Al-Hamamre Z, Al-Addous M, Alkasrawi M (2021) Pyrolysis of solid waste for bio-oil and char production in refugees' camp: a case study. Energies (Basel) 14. https://doi.org/10.3390/en14133861

6. Kumaravel ST, Murugesan A, Kumaravel A (2016) Tyre pyrolysis oil as an alternative fuel for diesel engines—a review
7. Yaqoob H, Teoh YH, Jamil MA, Gulzar M (2021) Potential of tire pyrolysis oil as an alternate fuel for diesel engines: a review
8. Zaiter MJ, Al-Asfoor IK, Almatari EK Salman ZT (2020) The production of pyrolytic oil from scrap tire tubes. Plant Arch 20
9. Haydary J, Susa D, Gelinger V, Čacho F (2016) Pyrolysis of automobile shredder residue in a laboratory scale screw type reactor. J Environ Chem Eng 4. https://doi.org/10.1016/j.jece.2015.12.038

Chapter 21
Experimental Study on the Impact Response of Carbon Fibre Reinforce Polymer Panels: Effect of Stacking Sequence

Muhammad Thaqib, Tajul Adli, Muhammad Azhad, Zainal Nazri Mohd Yusuf, and Muhammad Najib Abdul Hamid

Abstract This research focuses on the comprehensive characterization and evaluation of carbon fibre reinforced polymer (CFRP) composite laminates with diverse stacking sequences, specifically under low-velocity impact. The investigation applies Charpy impact and three-point bending tests to analyze the impact response and flexural behaviour. The study goes through the influence of stacking sequences on mechanical properties, including Young's modulus. The analysis validates experimental data and finite element analysis simulations for verification. Design 1 (using conventional stacking sequence) exhibits the highest ultimate tensile strength and brittleness, while design 2 (using alternating stacking sequence) presents intermediate strength. Design 3 (using an even stacking sequence) demonstrates unique behaviour with a "spring-like" quality. The research emphasizes the role of stacking sequences in CFRP behaviour, offering insights for car chassis applications. It

M. Thaqib · T. Adli (✉) · M. Azhad · M. N. A. Hamid
Engineering Section, Malaysian Spanish Institute Kulim Hi-Tech Park, Universiti Kuala Lumpur, Kulim, 09000 Kedah, Malaysia
e-mail: tajuladli@unikl.edu.my

M. Thaqib
e-mail: thaqib.zamnuri23@s.unikl.edu.my

M. Azhad
e-mail: Azhadmh@proton.com

M. N. A. Hamid
e-mail: mnajib@unikl.edu.my

Z. N. M. Yusuf
Mechanical Section, Malaysian Spanish Institute Kulim Hi-Tech Park, Universiti Kuala Lumpur, Kulim, 09000 Kedah, Malaysia
e-mail: zainalnazri@unikl.edu.my

M. Azhad
Proton Holdings Berhad, Persiaran Kuala Selangor, Seksyen 26, 40000 Shah Alam, Selangor, Malaysia

T. A. A. Razak et al. (eds.), *Progress in Engineering Technology VI*, Advanced Structured Materials 229, https://doi.org/10.1007/978-3-031-81517-1_21

acknowledges the limitations of measurement precision and underscores the importance of validating simulation results with experimental data. Integrating simulation tools like Abaqus proves crucial in optimizing CFRP product design and performance while maintaining alignment with real-world behaviour.

Keywords CFRP · Stacking sequences · Low-velocity impact · Brittleness · Charpy impact test

21.1 Introduction

Composite materials, particularly carbon fibre woven structures, have revolutionized numerous industries due to their exceptional properties [1, 2]. One promising rehabilitating and strengthening civil infrastructure technique involves carbon fibre reinforced polymer (CFRP) composites. A critical aspect within this domain is the arrangement of carbon fibre layers, known as stacking sequences, which significantly influence the mechanical behaviour of these composites [3].

Carbon fibre woven designs vary widely, characterized by the bundling of carbon fibres into tows classified by their filament count, such as 3 k, 6 k, 12 k, and 15 k [4–6]. The intricate weaving of carbon fibres within a confined space contributes to the remarkable strength of these materials [7–9]. Epoxies are commonly used in composites for their chemical resistance and thermo-mechanical properties, which can be tailored by adjusting their composition and curing kinetics [9, 10].

The chassis is a crucial structural element in vehicle design, supporting all components and enduring various forces [11–13]. The emergence of electric vehicles requires lightweight chassis as a trade-off for the heavy battery of the vehicle, and CFRP is suitable due to its weight efficiency. The performance of CFRP can be increased as different stacking sequences affect the tensile stress–strain relationship, a key indicator of toughness [14]. Charpy impact tests quantify toughness in Joules [15], aiming to identify the optimum stacking sequence suitable for car chassis applications.

This research aims to investigate optimal weaving patterns and stacking sequences to maximize the load-bearing capacity of carbon fibre slabs, particularly in car chassis applications. This study also acknowledges limitations such as the cost of CFRP materials, influenced by factors like the quality of polyacrylonitrile (PAN), rayon, and resin [16]. Additionally, natural curing methods are used due to the absence of specialized machinery, and the precision of data collected using analogue dials in the Charpy tester may affect measurement accuracy.

Fig. 21.1 1 × 1 twill

21.2 Methodology

Carbon fibre weave plays an important role in manipulating the CFRP outcome. The most common weave patterns are plain, twill, and satin. The weave pattern affects the properties of the carbon fibre fabric, such as its strength, stiffness, and density. This weave structure, where a tow strand traverses over a defined number of tows before passing beneath an equal number, establishes an over-and under pattern, crafting a distinctive "twill line"—a diagonal arrowhead motif. The enhanced inter-tow spacing in twill weaves translates to fewer crimps and diminished stress concentrations, setting it apart from plain weaves. Figure 21.1 shows the 1 × 1 twill. Figure 21.2 shows the quil orientation of carbon fibre filament.

The primary orientations are 0° and 45°, with the possibility of considering other orientations to optimize stacking sequences for improved structural responses under diverse stress conditions. The thickness average is ± 0.375 mm, necessitating 8 layers of carbon fibre to achieve a 3 mm thickness. The dimensions of the die-cast in the shape of a dogbone specimen are depicted in Fig. 21.3.

After determining the dimensions, 3 designs seem the most effective and can provide the best strain rate for the CFRP stacking sequence. Hence, Figs. 21.4, 21.5, and 21.6 are the chosen designs of the stacking sequence.

21.3 Results and Discussion

See Figs. 21.7, 21.8 and 21.9.

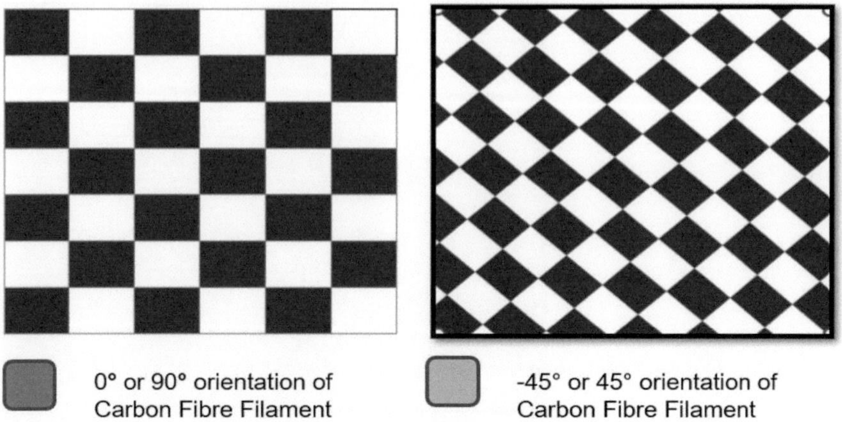

Fig. 21.2 Left shows the 90° quil orientation, and the right shows the 45° quil orientation

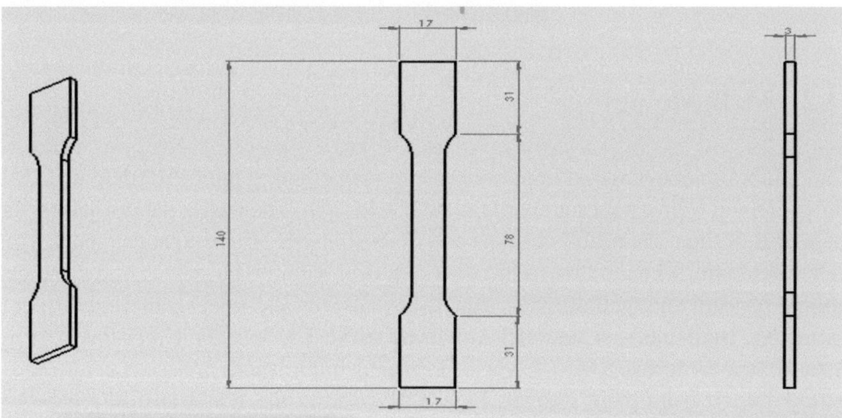

Fig. 21.3 Dimensions of the specimen

Fig. 21.4 Conventional stacking sequence

Fig. 21.5 Alternating stacking sequence

Fig. 21.6 Even stacking sequence

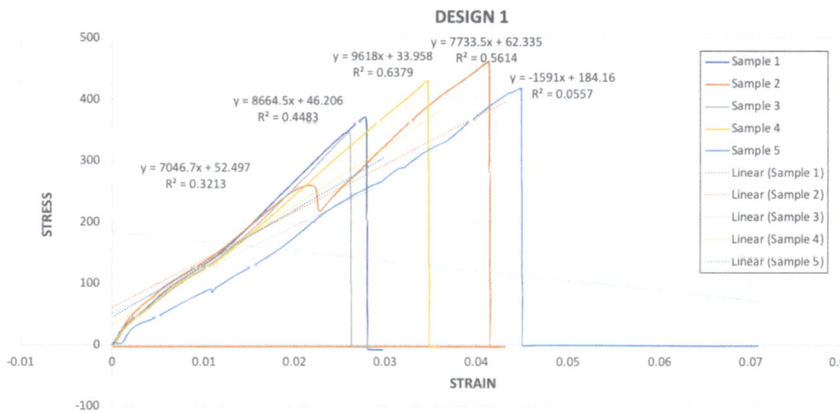

Fig. 21.7 Stress versus strain graph for design 1

21.3.1 Stress Versus Strain Graph Plot

A material's stress–strain relationship is illustrated through a stress–strain diagram or curve generated from mechanical tests where continuous stress and strain measurements are collected during load application. This curve graphically portrays the material's deformation response under tensile, compressive, or torsional forces. Stress–strain curves offer insights into crucial material characteristics, including the elastic and plastic regions, yield point, and ultimate tensile strength, contingent on the specific substance. In this project, data from UTM software was extracted and re-plotted in Microsoft Excel, yielding graphs for each design as illustrated in Figs. 21.7, 21.8 and 21.9.

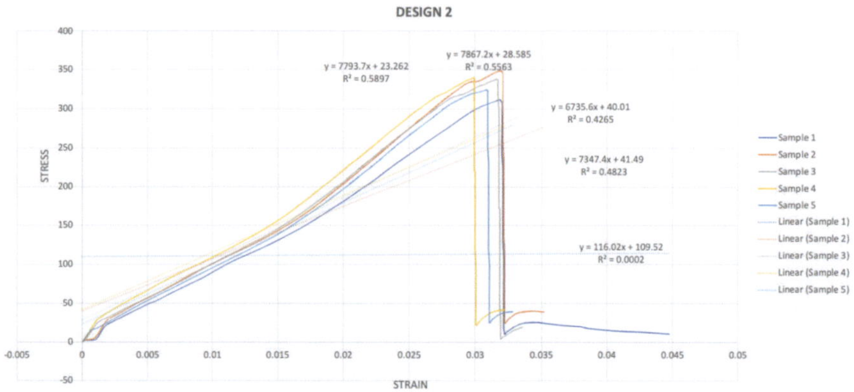

Fig. 21.8 Stress versus strain graph for design 2

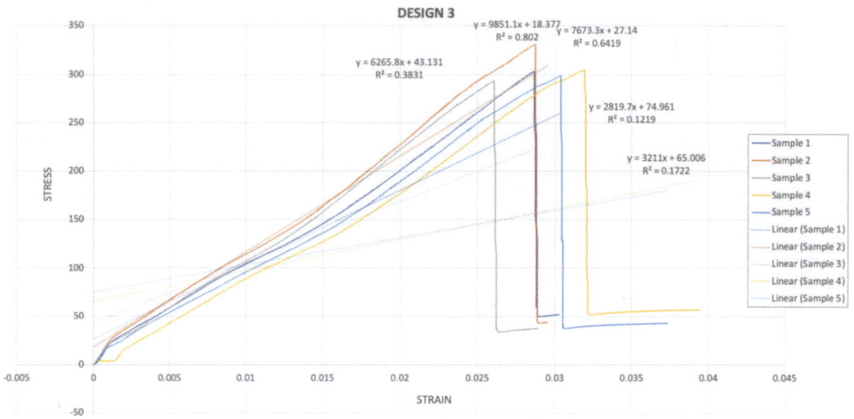

Fig. 21.9 Stress versus strain graph for design 3

21.3.2 Elastic/Young's Modulus of Carbon Fibre

Young's modulus, also termed the modulus of elasticity, characterizes a material's capacity to withstand compression or elongation relative to its length. The provided figure was generated using raw experimental data, focusing on obtaining an accurate elastic modulus. To achieve this, the maximum tensile stress from each specimen was selected and plotted, with the modulus derivable from the stress–strain graph's gradient. Consequently, Figs. 21.10, 21.11 and 21.12 are the graphs; Table 21.1 was created to illustrate these findings.

Table 21.1 summarizes the experimental results for each specimen's Young's modulus. A corresponding finite element analysis (FEA) simulation was conducted using identical parameters, focusing on Young's modulus and maximum force in the Y direction to validate these findings. Initial data from the universal tensile

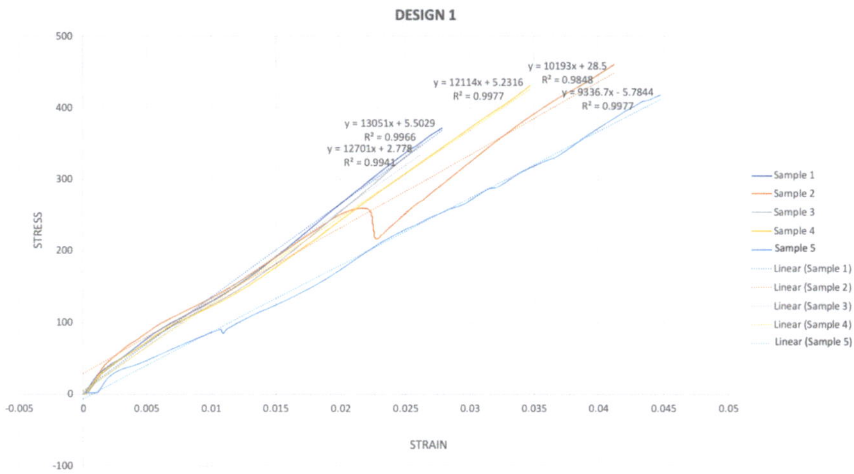

Fig. 21.10 Trim stress versus strain graph for design 1

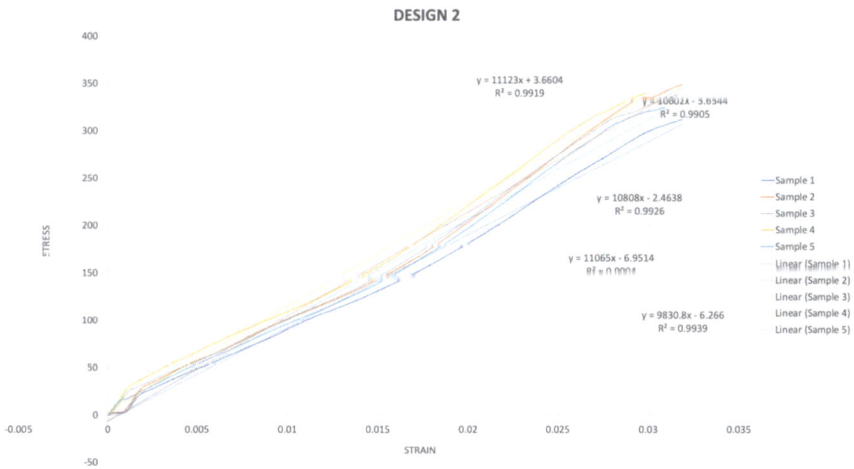

Fig. 21.11 Trim stress versus strain graph for design 2

machine (UTM) was processed in Excel to refine stress–strain curves and determine the ultimate tensile strength (UTS) values for specific designs, providing insights into CFRP's orientation effects.

Design 1, oriented at 0°, exhibited the highest UTS, indicating its ability to withstand high-stress levels but demonstrating brittle behaviour upon failure. Design 2, featuring alternating 0° and 45° orientations, displayed intermediate strength and fracture characteristics similar to design 1. In contrast, design 3, with equal layers of 0° and 45° orientations, showed lower UTS but exhibited a unique behaviour: the

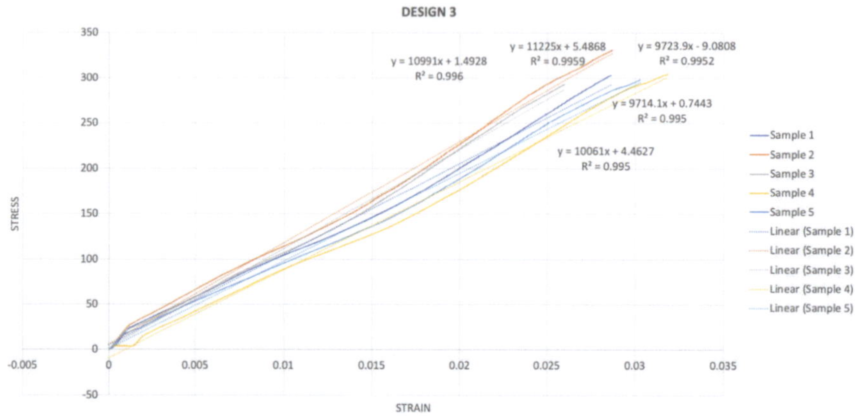

Fig. 21.12 Trim stress versus strain graph for design 3

Table 21.1 The Young's modulus data

Design	Elastic modulus (MPa)				
	Sample 1	Sample 2	Sample 3	Sample 4	Sample 5
1	13,051	10,193	12,701	12,114	9336.7
2	9830.8	11,065	10,808	11,123	10,602
3	10,061	11,225	10,991	9723.9	9714.1

45° layers tore under stress while the 0° layers demonstrated resilient deformation, which could be advantageous in low-impact scenarios.

The simulation generated stress–strain curves illustrating stress distribution patterns, validated through mean absolute percentage error calculations against experimental data. This affirmed that stacking sequence orientations significantly influence carbon fibre behaviour, impacting material performance under different loading conditions.

This study contributes empirical data on CFRP behaviour and explores its practical implications in car chassis applications. Design recommendations vary: Design 1 is suitable for applications requiring high stiffness and resistance to deformation, while design 3's characteristics may be advantageous in scenarios necessitating energy absorption and structural integrity preservation. Ultimately, the optimal design choice hinges on specific application requirements, aligning with the observed behaviour of each stacking sequence design.

21.4 Conclusion

In conclusion, this study integrated experimental findings with finite element analysis simulations to explore the behaviour of carbon fibre reinforced polymers under varied stacking sequences. Excel software facilitated initial data analysis, focusing on ultimate tensile strength values and Young modulus calculations from stress–strain curves. Design 1 exhibited high UTS but brittleness. Design 2 showed intermediate strength with mixed orientations, and design 3 demonstrated resilience despite lower UTS. Abaqus software validated simulation accuracy, predicting stress–strain curves and damage evolution for CFRP design optimization. Experimental validation remains crucial for ensuring reliability in simulated material behaviour, highlighting Abaqus' role in enhancing CFRP performance and design.

References

1. Batuwitage C, Fawzia S, Thambiratnam D, Al-Mahaidi R (2017) Durability of CFRP strengthened steel plate double-strap joints in accelerated corrosion environments. Compos Struct 160:1287–1298
2. Tafsirojjaman T, Fawzia S, Thambiratnam DP, Zhao X-L (2020) Study on the cyclic bending behaviour of CFRP strengthened full-scale CHS members. Structures 28:741–756
3. Naveen R, Kumar M, Matahn A, Dhushyanath D (2021) Investigation on the effect of stacking sequence on mechanical properties of a basalt and carbon fiber hybrid composite. J Eng Res 9:1–14
4. Wahidah N et al (2022) Mechanical and physical properties of recycled-carbon-fiber-reinforced polylactide fused deposition modelling filament. Materials (Basel) 15(1):1–16
5. Lupone F, Padovano E, Venezia C, Badini C (2022) Experimental characterization and modeling of 3D printed continuous carbon fibers composites with different fiber orientation produced by FFF process. Polymers (Basel) 14(3)
6. Goh GD et al (2018) Characterization of mechanical properties and fracture mode of additively manufactured carbon fiber and glass fiber reinforced thermoplastics. Mater Des 137:79–89
7. Awla NJ, Khdir YK (2023) Experimental study of mechanical properties of unidirectional woven carbon fiber and fiber glass—epoxy composites. Al-Nahrain J Eng Sci 26(1):7–13
8. Kochov A, Spasovski M, Serafimovski F (2022) Tensile strength of carbon fiber. Teh Glas 16(4):537–541
9. Thakre PR et al (2011) Investigation of the effect of single wall carbon nanotubes on interlaminar fracture toughness of woven carbon fiber—epoxy composites
10. Njuguna J, Pielichowski K, Fan J (2012) Polymer nanocomposites for aerospace applications. Woodhead Publishing Limited
11. Zaenuri A, Sunardi A, Kasum (2022) Design and manufacture of electric car chasis frame for two passengers with solidworks modeling 2020. J Glob Eng Res Sci 1(1):12–16
12. Alyona L et al (2022) Determining features of the stressed state of a passenger car frame with an energy-absorbing material in the girder beam. Eastern-European J Enterp Technol 5(7–119):44–53
13. Živile Decker et al (2023) Analysis of the vehicle chassis axle fractures. Materials (Basel) 16(2)
14. Jianwu Zhou et al (2019) Low-velocity impact behavior and residual tensile strength of CFRP laminates. Compos Part B Eng 161:300–313
15. Caminero MA, Rodríguez GP, Muñoz V (2016) Effect of stacking sequence on Charpy impact and flexural damage behavior of composite laminates. Compos Struct 136:345–357

16. Hamid K et al (2020) PAN precursor fabrication, applications and thermal stabilization process in carbon fiber production: experimental and mathematical modelling. Prog Mater Sci 107:100575

Chapter 22
Vibration Attenuation of Unbalanced Motor Using a Tuneable Dynamic Vibration Absorber

Fazidah Saad, Muhammad Najib Abdul Hamid, Zainal Nazri Mohd Yusuf, and Ahmad Zhafran Ahmad Mazlan

Abstract Generally, a moving system will typically produce vibrations when it is moving in rectilinear or rotating and it is common for the unbalanced motor to produce high vibration intensity when it is operated. This vibration is very harmful to the motor's lifespan and to its operations. Therefore, it is very important to control and reduce the vibrations produced by the unbalanced motor during operation. One of the popular methods of doing this is using dynamic vibration absorbers (DVAs). DVA is an efficient control device that is connected to a vibrating system to reduce undesired vibrations by equalizing its excitation frequency to nearly coincide with the natural frequency of the primary system. This can be achieved by tuning the DVA parameters to the optimal value. This research is aimed at designing and developing a tuneable dynamic vibration absorber (TDVA) to reduce the unbalanced motor vibration by optimizing the DVA parameters. The optimal DVA parameters have been computed directly using dynamic properties of the primary system by assuming the unbalanced motor to be the single degree of freedom vibration system. The results of this study indicate that the total vibration reduction gained by the unbalanced motor when using DVA is about 96.85% where the initial amplitude of the motor before and after the DVA attachment is 4.13 m/s^2 and 0.13 m/s^2. This has shown that the designed TDVA is operating as anticipated because it was able to successfully reduce the amplitude.

F. Saad (✉) · M. N. A. Hamid · Z. N. M. Yusuf
Mechanical and Engineering Section, Malaysian Spanish Institute Kulim Hi-Tech Park, Universiti Kuala Lumpur, Kulim, 09000 Kedah, Malaysia
e-mail: fazidah@unikl.edu.my

M. N. A. Hamid
e-mail: mnajib@unikl.edu.my

Z. N. M. Yusuf
e-mail: zainalnazri@unikl.edu.my

A. Z. A. Mazlan
School of Mechanical Engineering, Universiti Sains Malaysia, Nibong Tebal, 14300 Pulau Pinang, Malaysia
e-mail: zhafran@usm.edu.my

T. A. A. Razak et al. (eds.), *Progress in Engineering Technology VI*, Advanced Structured Materials 229, https://doi.org/10.1007/978-3-031-81517-1_22

Keywords Unbalanced motor · Tuneable dynamic vibration absorber · TDVA · Mass ratio

22.1 Introduction

Any rotating imbalance such as the unbalanced motor will create vibration. This vibration diminishes the lifespan of the motor and in the worst scenario, it can make the motor break down. Reducing vibration or dynamically balancing the spinning assembly are the best ways to avoid this. While there are numerous ways to manage and lessen vibration, one of the most effective ones is to use a dynamic vibration absorber (DVA).

Dynamic vibration absorbers (DVAs), also known as tuned mass dampers (TMDs), are mechanical add-ons that are made up of mass, stiffness, and damping elements that can absorb the vibration energy of a structure which usually referred to primary system when they are connected. Frahm [1] presented the earliest conception of a vibration absorber. With the aid of DVA, vibrations can be transferred from the primary structure to the secondary structure which is the DVA itself so that the excessive vibration energy input to the primary structure is subsequently released through the secondary structure's vibration. This is done by equaling the DVA natural frequency to the frequency of the primary system.

DVAs are available in a variety of configurations that are designed to attenuate either rectilinear or angular motion. The simplest configuration is made up of a single mass connected to the main system by a linear spring. This arrangement is always referred as undamped DVA. The main problem with this configuration is to tune the DVA to attenuate harmonic vibrations, the mass and stiffness of the undamped DVA need to be selected in accordance with the value of the excitation frequency or else it may lose its effectiveness [2].

Recently, researchers have shown an increased interest in the wide band frequency control function of a DVA because of its practicality and there are numerous studies attempted to propose the optimal design of DVA parameters. For instance, there is a study to optimize the tuning ratio of a DVA [3], optimizing the DVA parameters under seismic excitation [4], tuning the natural frequency of a distributed DVA around the primary system [3, 5, 6] and a lot more. Besides that, the optimization of the DVA parameters such as mass, stiffness, and damping also has been widely investigated [7] but there is still an issue with the location of DVA attachment to the primary system as it will definitely affect the characteristics of the whole system as well [8]. There are three factors that can be chosen in pre-tuning the absorber's performance if the primary structure is assumed to be presented by a SDOF system and they are: the frequency ratio, the damping ratio, and the absorber mass. For the reason of deliberation, the mass ratio between the primary system and the DVA is normally taken as a constant value in the pre-design stage [9]. According to Brennan [10], the DVA can be tuned to be a resonant device and can regulate the structure if it is placed at correct location and it can offer effective control in a wide frequency range.

Therefore, when controlling a wide frequency bandwidth, the optimal locations of the absorbers should be identified after knowing the trade-off between the targeted mode and other modes in the system [8].

The goal of this study is to design and develop a tuneable dynamic vibration absorber (TDVA) that can attenuate the unbalanced motor vibration by restraining the mass ratio of the system to the maximum allowable value for its effective operation. For the TDVA to successfully absorb the vibrations at a wide range of the frequency band, the TDVA placement is tuned to be at the ideal location.

22.1.1 Experimental Modal Analysis and Vibration Measurements of the Primary System

Experimental modal analysis (EMA) is one of the processes used to find the natural frequency and vibration properties of a structure. It is important to understand the vibrations dynamic properties of a structure for the purpose of designing or optimizing it.

In this study, the EMA is performed by using the LMS TestLab software that is installed on a computer that is connected to a DAQ and the DAQ is coupled to an impact hammer and accelerometer that is attached to the unbalanced motor base. The accelerometer is attached to the unbalanced motor base plate to measure the natural frequency of the unbalanced motor during the impact given by the impact hammer. From EMA of the unbalanced motor, the targeted frequency to be attenuated in this system is identified as its natural frequency.

Once the natural frequency of the primary system was identified, the data were used to design the tunable dynamic vibration absorber parameters for the unbalanced motor.

22.1.2 Design of Tuneable Dynamic Vibration Absorber

A vibration absorber is a second spring-mass system added to the primary mass and it is designed to absorb the input disturbance by shifting the motion to the newly added mass which is the absorber mass. The designed structures of the unbalanced motor and the TDVA are shown in Fig. 22.1 and schematically in Fig. 22.2, in which the TDVA is installed on the shaft that is attached underneath the unbalanced motor. The TDVA consists of a steel mass that is attached to the steel rod at a certain length. The stiffness of the TDVA depends on the stiffness of the steel rod that retains the absorber mass.

Figure 22.3 shows the flowchart of the TDVA design process. It started with identifying the natural frequency of the primary system which is the unbalanced

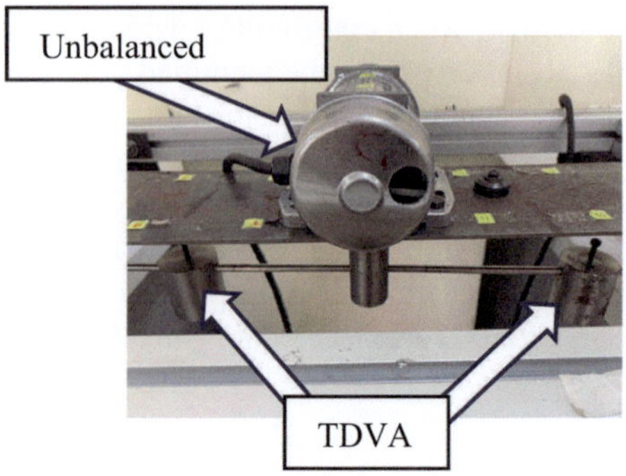

Fig. 22.1 Unbalanced motor with TDVA attached underneath

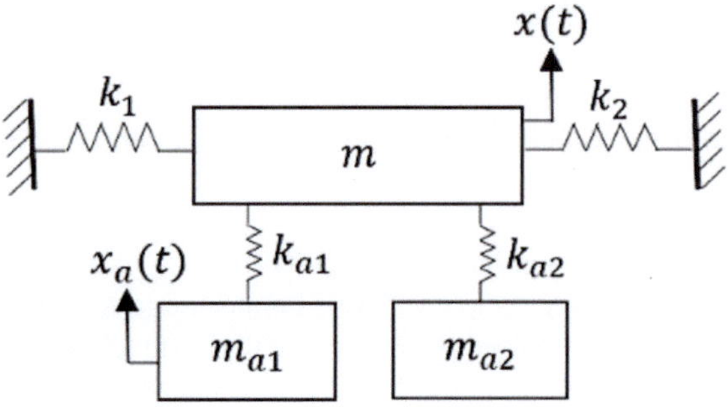

Fig. 22.2 Schematic diagram of the unbalanced motor and TDVA

motor by EMA. The primary system will then be measured to continue calculating all the parameters required for the TDVA.

The frequency ratio β is the ratio between the absorber's natural frequency, ω_{a} and the primary system's natural frequency, ω while the mass ratio, μ is the ratio between the absorber mass, m_{a} and primary mass m.

If the frequency ratio, β of the system is fixed so that the natural frequency of the primary system and the absorber is the same, the effective range of mass ratio curve is as shown in Fig. 22.4.

From Fig. 22.4, if the mass ratio, μ is too high, the frequencies will split further apart ($> \omega_2$), and if the ratio is too small, the DVA will not be able to withstand much variation in driving frequency ($< \omega_1$), which would reduce the DVA's effectiveness. It has been suggested that the vibration intensity is decreasing as the mass ratio rises [9].

Therefore, the rule of thumb used for the effective range of mass ratio, μ for the dynamic vibration absorber needs to be in the range of 0.05 to 0.25 [11]. Consequently, the mass ratio, μ chosen for this study is based on the maximum value in the effective range, which is 0.25.

The design of TDVA begins with the measurement of the unbalanced motor mass, m which is 3.975 kg. The absorber mass m_a was calculated to be 0.99375 kg an in Eq. (22.1).

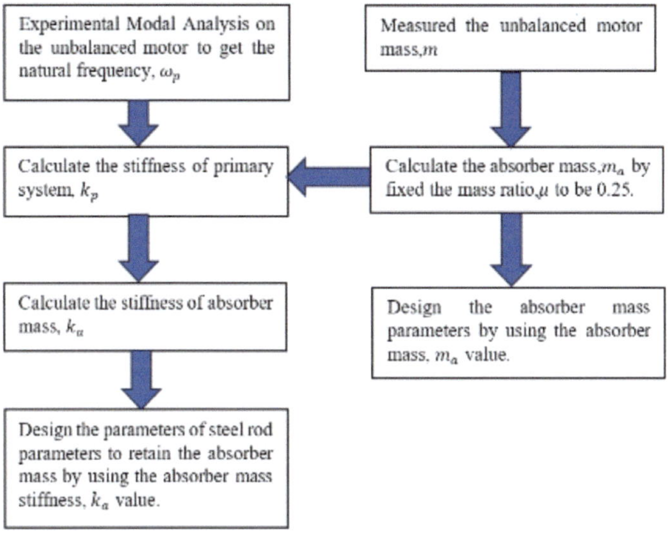

Fig. 22.3 Flowchart of TDVA design

Fig. 22.4 Effective range of mass ratio curve [11]

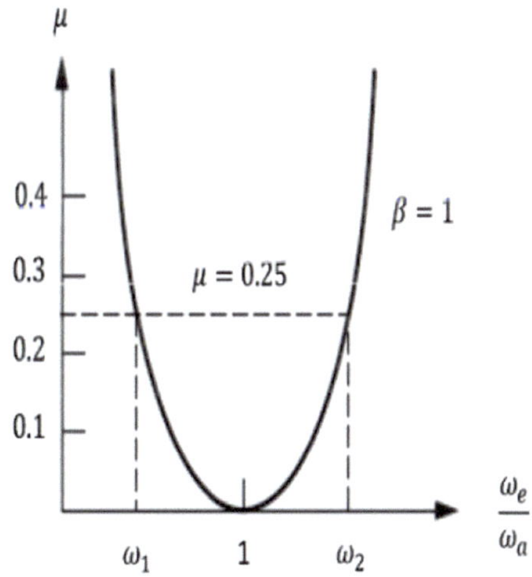

$$m_a = \mu m = 0.25 \times 3.975 \text{ kg} = 0.99375 \text{ kg} \tag{22.1}$$

However, by considering there will be two absorber masses that are equally located on both sides of the primary system, therefore, the absorber mass is divided by two.

Then the stiffness of the primary system is calculated as in Eq. (22.2) by using the value of unbalanced motor natural frequency from EMA which is 19.57 Hz.

$$k_p = (\omega_p)^2 \times m_p = (2 \times \pi \times 19.57)^2 \times 3.975 = 60.10 \text{ kN/m} \tag{22.2}$$

Table 22.1 Steel rod parameters [12, 13]

Young Modulus for steel, E	190 GPa
Density of steel, ρ	7800 kg/m^3
Steel rod radius, r_r	0.003 m
Moment of inertia for a rod, I	$I = \frac{1}{4}\pi r^4 = 6.36173e^{-11}$ m^4
Absorber volume, V_a	$V_\mathrm{a} = \frac{m_\mathrm{a}}{\rho} = \frac{0.49688}{7800} = 6.37e^{-5}$ m^3

The focus of the TDVA is to concentrate at the highest frequency of the primary system, which is at the resonance, therefore, the natural frequency of the absorber is assumed to be the same as the natural frequency of the primary system as well. Therefore, $\omega_\mathrm{a} = \omega_\mathrm{p} = 19.57$ Hz.

Then, the stiffness of the absorber is calculated as in Eq. (22.3).

$$k_\mathrm{a} = (\omega_\mathrm{a})^2 \times m_\mathrm{a} = (2 \times \pi \times 19.57)^2 \times 0.49688 = 7.513 \text{ kN/m} \tag{22.3}$$

In this study, the stiff component of the absorber is made of a steel rod, and the absorber mass is made of solid cylindrical steel. The computation of the steel rod design is carried out utilizing the mechanical properties of steel. The calculations are as in Table 22.1.

Equation (22.4) can be used to compute the absorber's parameters when the volume of the absorber is known.

$$V_\mathrm{a} = \pi r^2 h = \pi \times (0.02 \text{ m})^2 \times 0.0\ 5\text{m} = 6.2382e^{-5} \text{ m}^3 \tag{22.4}$$

As a result, the solid cylindrical steel's specifications for the absorber mass are as follows; radius, r is 0.02 m and height, h is 0.05 m.

22.2 Tuneable Dynamic Vibration Absorber Model Construction and Validation

In this study, the stiffness of the absorber mass is affected by the length of the steel rod used to retain the absorber mass, which allows the absorber vibration to be attenuated. Therefore, the length of the steel rod to be used to retain the absorber mass is calculated as follows: considering the steel rod attached to the unbalanced motor as a cantilever beam with a point load at its free end, therefore, the maximum beam deflection is as shown in Eq. (22.5).

$$\delta = \frac{FL^3}{3EI} \tag{22.5}$$

where δ is the maximum deflection in m, F is the force applied in N, L is the length in m, E is the Young's modulus in Pa and I is the moment of inertia for a rod in m^3.

Subsequently, Eq. (22.6) shows the Hooke's law that states the applied force, F is equal to the stiffness,k times the displacement or the deflection, δ. In this case, the stiffness used is the absorber stiffness, k_a.

$$F = k_a \delta \tag{22.6}$$

By substituting Eqs. (22.5) and (22.6), the length of the steel rod is determined by Eq. (22.7).

$$L = \sqrt[3]{\frac{3EI}{k_a}} = \sqrt[3]{\frac{3 \times 190e^9 \times 6.36173e^{-11}}{7.513e^3}} = 0.1686 \text{ m} \tag{22.7}$$

The absorber model is then built using solid cylindrical steel utilizing the previously estimated parameters. The absorber mass was then secured to the steel rod that was connected to the unbalanced motor at 0.1686 m away from the motor's center. The entire unbalanced motor system with the TDVA attached is then operated at the targeted vibration for this study which is at the resonance frequency of 19.57 Hz to observe the outcomes for results validation purposes.

22.3 Results and Discussion

Based on the information obtained from EMA of the unbalanced motor, the TDVA model was calculated and created. The TDVA was fabricated, and an experimental investigation was conducted on the entire system operating at the intended frequency which is at the resonance with the TDVA attached to the unbalanced motor. The results of the experiment indicate the vibration attenuation by using TDVA.

22.3.1 *Spectral Analysis During the Unbalanced Motor Operation at Resonance Frequency*

Spectral analysis by using the LMS Software was performed at the unbalanced motor during operation at the resonance frequency which is at 19.50 Hz. From the analysis, the amplitude of the unbalanced motor was recorded to be 4.13 m/s^2 as in Fig. 22.5. This amplitude appears to be high and can be reduced by implementing the TDVA to the system.

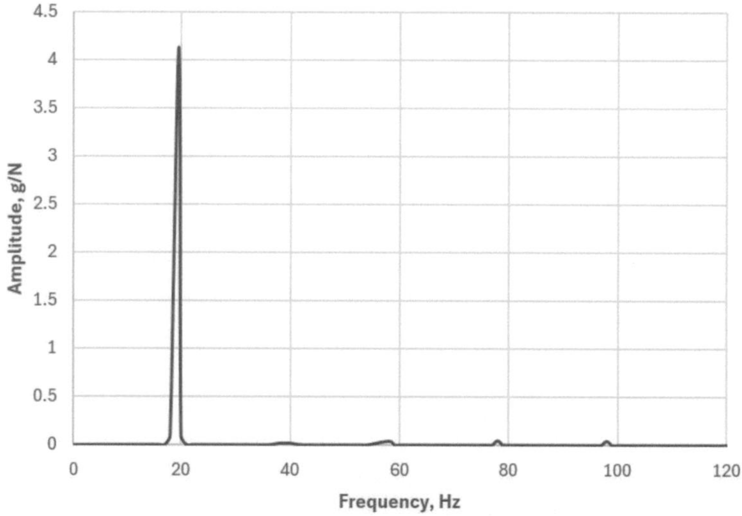

Fig. 22.5 Spectral analysis of the unbalanced motor without DVA

22.3.2 Spectral Analysis During the Unbalanced Motor Operated at Resonance Frequency with TDVA Attached

In order to diminish the vibration at the resonance of the primary system, the designated TDVA was fabricated and attached to the unbalanced motor. After that, spectral analysis was conducted on the entire system with the TDVA attached to it in order to verify the system's vibration reduction.

EMA When TDVA Attached at the Calculated Length

Firstly, the TDVA was attached to the steel rod at the designated length as per calculation made earlier which is at 0.1686 m from the center at both sides of the steel rod and the system has undergone EMA to verify the natural frequency that needs to remain the same as the primary system at 19.57 Hz. But the results show that at the length of 0.1686 m, the natural frequency had deviated to 20.02 Hz as shown in Fig. 22.6. Therefore, the length of the absorber mass attached to the steel rod is tuned until the natural frequency of the whole system is the same as the primary.

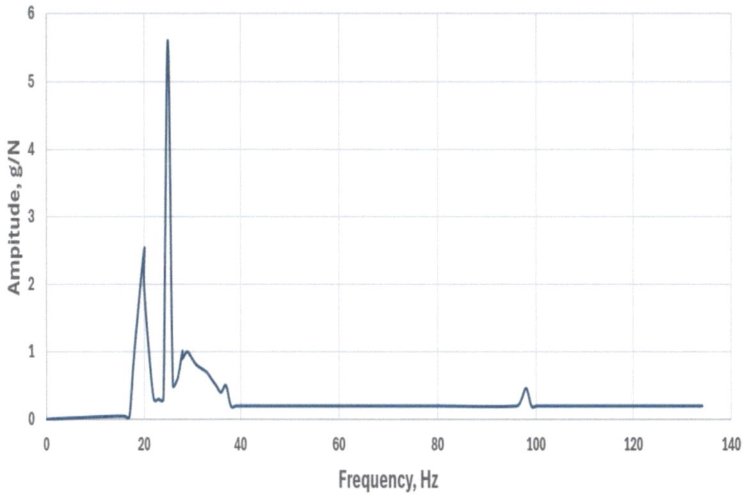

Fig. 22.6 EMA of the unbalanced motor with DVA attachment at 0.1686 m (calculated length) from the center of primary system

EMA and Spectral Analysis When TDVA Attached at the Optimized Length

The length of the steel rod for the TDVA attachment has been optimized manually by doing EMA at different locations of steel rod length. It appears from the EMA that the adjusted locations are at a distance of 0.1486 m from the center of the unbalanced motor. This can be seen in Fig. 22.7.

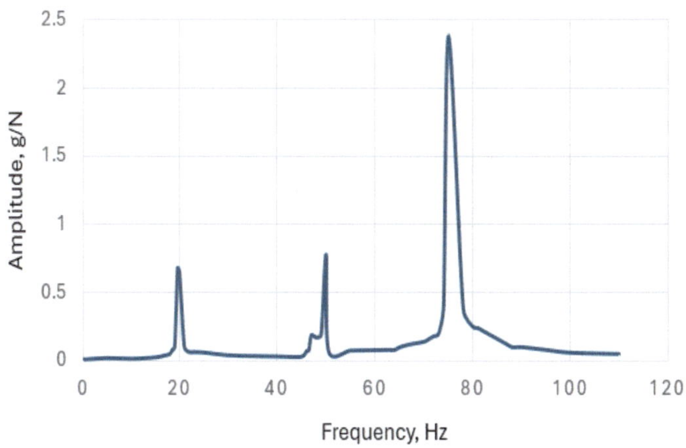

Fig. 22.7 EMA of the unbalanced motor with DVA attachment at 0.1486 m (optimized length) from the center of primary system

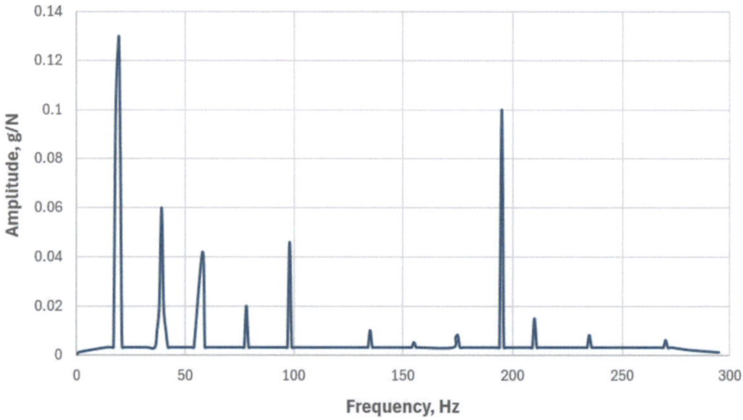

Fig. 22.8 Spectral analysis of the unbalanced motor with DVA attachment at 0.1486 m optimized length) from the center of primary system

Following the confirmation of the attachment length, a spectral analysis of the entire system was performed to ascertain the vibration reduction when it was operating at the resonance frequency. From Fig. 22.8, the result of the spectral analysis shows there is an enormous vibration reduction to the system when applying the TDVA with the amplitude is indicating at 0.13 m/s^2 if compared to 4.13 m/s^2 before. This has shown a prodigious vibration diminution of 96.85%.

22.4 Conclusion

The goal of this study is to develop a TDVA that can reduce the vibration caused by unbalanced motors by confining the mass ratio to the upper permissible limit for its effective operating range. The results of this study show that the unbalanced motor's maximum amplitude gain occurs when it is operating at its natural frequency, which is 4.13 m/s^2. The amplitude was drastically reduced by 96.85% to just 0.13 m/s^2 after the TDVA was connected to the motor. This has demonstrated that the intended TDVA is operational correctly because it has succeeded in reducing such a significant amount of amplitude.

The dimensions of the TDVA itself and the placement of its attachment to the steel rod are the two important design considerations to reduce the vibration at the primary system because they have a significant effect on the absorber stiffness. According to the calculation, the TDVA must be retained at 0.1686 m from the unbalanced motor's center on both sides of the steel rod, albeit during the testing, the distance was really at 0.1486 m. This indicates that there was a minor issue with the manufactured TDVA, which is most likely caused by the volume of the TDVA itself. The percentage of error for the position of the TDVA is roughly 11.86%. But in the end, the TDVA is sufficient to lessen the unbalanced motor's vibration.

In conclusion, the current study has made some progress in improving the understanding of how the DVA's design can be adjusted to have the necessary stiffness of the absorber in order to maximize vibration absorption from the primary system.

References

1. Frahm H (1911) Device for damping vibrations of bodies
2. Steffen V, Rade D (2001) Absorbers Vibration Encycl Vib 9–26 https://doi.org/10.1006/rwvb.2001.0176
3. Beskhyroun S, Wegner LD, Sparling BF (2011) Integral resonant control scheme for cancelling human-induced vibrations in light-weight pedestrian structures. Struct Control Heal Monit n/a-n/a. https://doi.org/10.1002/stc
4. Bekdaş G, Nigdeli SM (2011) Estimating optimum parameters of tuned mass dampers using harmony search. Eng Struct 33:2716–2723. https://doi.org/10.1016/j.engstruct.2011.05.024
5. Alshabatat NT (2012) Optimizing the natural frequencies of beams via notch stamping. Res J Appl Sci Eng Technol 4:2030–2035
6. Hu ZK, Wang M, Zan T (2013) Dynamic vibration absorber design to suppress boring chatter: absorber parameters identification based on modal correlation. Adv Mater Res 753–755:1816–1820. https://doi.org/10.4028/www.scientific.net/AMR.753-755.1816
7. Zilletti M, Elliott SJ, Rustighi E (2012) Optimisation of dynamic vibration absorbers to minimise kinetic energy and maximise internal power dissipation. J Sound Vib 331:4093–4100. https://doi.org/10.1016/j.jsv.2012.04.023
8. Yang C, Li D, Cheng L (2011) Dynamic vibration absorbers for vibration control within a frequency band. J Sound Vib 330:1582–1598. https://doi.org/10.1016/j.jsv.2010.10.018
9. Younes MF (2017) Optimal design of dynamic vibration absorber for rolling systems. J Eng Appl Sci 64:307–324. https://doi.org/10.21608/amme.2018.34728
10. Brennan MJ, Dayou J (2000) Global control of vibration using a tunable vibration neutralizer. J Sound Vib 232:585–600. https://doi.org/10.1006/jsvi.1999.2757
11. Mazlan AZA, Ripin ZM (2015) Active vibration control to attenuate hand-arm vibration from orbital sander: a mathematical model approach. Indian. J Sci Technol 8. https://doi.org/10.17485/ijst/2015/v8i30/85942
12. Is W, Of M What is Young's Modulus or Modulus of Elasticity?—ExtruDesign. https://extrudesign.com/youngs-modulus-modulus-of-elasticity/
13. Martin M (2004) Density of steel. Hypertextbook.Com 3:1–1

Chapter 23
The Study of Mechanical Properties of Palm Oil Kernel Shell Composite Using the Taguchi Method

Mohamad Farhan Bin Mohd Fauzi, Mohamad Sazali Bin Said, Mohd Shahrizan Bin Yusoff, Shahruzaman Bin Sulaiman, and Syed Ahmad Faiz Syed Ahmad

Abstract Palm oil mills have emerged as significant contributors to industrial pollution in recent years. As concerns over environmental issues and global warming continue to gain traction in society, research efforts focusing on palm kernel shell reinforcement composites have intensified. The use of eco-friendly materials serves as a primary motivation for researchers in this field. In order to investigate the mechanical properties of palm kernel shell composites, the Taguchi method was employed. Palm oil kernel shell weight percentage and tensile test speed significantly affected tensile strength, modulus of elasticity, and elongation at break. In line with the ASTM D638-14 standard, modifications were made to the mold, and several specimens were subsequently developed. Tensile specimens, comprising 30%, 40%, and 50% palm kernel shell composites, were carefully fabricated. The mechanical properties of the palm kernel shell composites were evaluated using the universal testing machine and the Taguchi method orthogonal array L9. The findings from the Taguchi analysis revealed that a weight percentage of 40% was recommended. The tensile strength was measured to be 22.06 N/mm^2, the modulus of elasticity was determined to be 2216.82 N/mm^2, and the elongation at break was 1.87%. Notably, the palm kernel shell composites' mechanical properties were independent of the weight proportion.

M. F. B. M. Fauzi · M. S. B. Said (✉) · M. S. B. Yusoff · S. B. Sulaiman · S. A. F. S. Ahmad
Malaysian Spanish Institute Kulim Hi-Tech Park, Universiti Kuala Lumpur, Kulim, 09000 Kedah, Malaysia
e-mail: msazali@unikl.edu.my

M. F. B. M. Fauzi
e-mail: farhan.fauzi13@s.unikl.edu.my

M. S. B. Yusoff
e-mail: mshahrizan@unikl.edu.my

S. B. Sulaiman
e-mail: shahruzaman@unikl.edu.my

S. A. F. S. Ahmad
e-mail: syedahmadfaiz@unikl.edu.my

T. A. A. Razak et al. (eds.), *Progress in Engineering Technology VI*, Advanced Structured Materials 229, https://doi.org/10.1007/978-3-031-81517-1_23

This observation suggests that palm kernel shell composites can potentially replace existing products with biodegradable alternatives, thereby mitigating pollution from costly decomposition processes.

Keywords Mechanical properties · Palm kernel shell · Epoxy resin · Natural fiber composite · Taguchi method

23.1 Introduction

The escalating global demand for sustainable and environmentally friendly materials has generated a surging interest in the utilization of agricultural byproducts in diverse industrial applications. Among these byproducts is the palm oil kernel shell (PKS), a fibrous residue produced during the palm oil milling process. Palm oil production is a significant industry in Malaysia and Indonesia, where the climate is highly conducive to oil palm tree cultivation. While the palm oil industry contributes significantly to the economic progress of these nations, it also generates substantial waste, including PKS. Historically, PKS has been utilized as a low-cost fuel source due to its high calorific value; however, exploring its potential as a composite material is imperative to enhance its economic and environmental value.

Despite the extensive research on incorporating natural fibers and agricultural residues into composite materials, there is a notable gap in the existing literature concerning the optimization of the mechanical properties of PKS composites. The mechanical properties of composite materials play a pivotal role in determining their suitability for various industrial applications, such as in the automotive, construction, and aerospace sectors. Nevertheless, optimizing these properties often necessitates intricate experimental designs that can be both time-consuming and resource-intensive.

To address this gap, the present study employs the Taguchi method, a robust statistical tool developed to optimize process parameters with minimal experimentation. The Taguchi method is well-suited for studies involving multiple variables, as it efficiently identifies the most influential factors and their optimal levels. By employing the Taguchi method to investigate PKS composites, this research aims to systematically examine the effects of different processing parameters, specifically the weight of composite particles, on key mechanical properties, including tensile strength, flexural strength, and impact resistance.

The outcomes of this study will contribute to the broader field of sustainable material development by offering valuable insights into the effective utilization of PKS as a composite material. Furthermore, the optimized mechanical properties of PKS composites could facilitate their application in various high-performance industrial sectors, thereby supporting economic and environmental sustainability.

23.2 Literature

Palm oil kernel shell is an abundant byproduct of the palm oil industry, particularly prevalent in palm oil-producing countries such as Malaysia and Indonesia. Traditionally, PKS has been utilized as a low-cost fuel due to its high calorific value and ease of transport. However, in recent years, there has been a growing interest in exploring the potential of PKS as a composite material, driven by the need for more sustainable and eco-friendly alternatives to conventional materials. Research has shown that PKS possesses unique properties, such as low density and high hardness, which make it a promising candidate for various applications, including construction, automotive, and packaging industries.

Several studies have investigated the incorporation of PKS into polymer matrices to create composite materials with enhanced mechanical properties. For instance, Adeniyi et al. [1] reported that the addition of PKS to epoxy resin improved the tensile strength and modulus of the resulting composite, making it suitable for lightweight construction materials. Similarly, studies Anyaoha et al. [2] demonstrated the potential of PKS as a reinforcement material in thermoplastic composites, where it contributed to increased stiffness and impact resistance. Despite these promising findings, the mechanical properties of PKS composites remain highly dependent on factors such as particle size, weight fraction, and processing conditions, indicating a need for further optimization.

23.2.1 The Taguchi Method in Composite Material Research

The Taguchi method, developed by Genichi Taguchi in the 1950s [3], is a robust statistical approach widely used for optimizing manufacturing and material science process parameters. The method employs orthogonal arrays to systematically explore the effects of multiple variables with a minimal number of experiments, thereby reducing time and resource consumption. The Taguchi method is particularly effective in identifying the most influential factors and their optimal levels, making it an invaluable tool in studies where multiple parameters must be considered.

In composite material research, the Taguchi method has been applied to optimize various properties, including mechanical strength, thermal stability, and wear resistance. For example, Prasad et al. [4] utilized the Taguchi method to optimize the tensile strength of natural fiber-reinforced composites, identifying fiber length and orientation as key factors. Similarly, Biswas and Satapathy [5] applied the Taguchi method to evaluate particulate-reinforced composites' wear behavior, achieving significant wear resistance improvements by optimizing the particle size and content.

Despite the widespread application of the Taguchi method in composite research, its use in optimizing PKS composites remains limited. Most existing studies have focused on traditional trial-and-error approaches, which are less efficient and often fail to capture the complex interactions between variables. This research aims to address this gap by applying the Taguchi method to systematically investigate the effects of weight and tensile test speed parameters on the mechanical properties of PKS composites.

While the potential of PKS as a composite material has been recognized, there is a lack of comprehensive studies that apply advanced statistical methods, such as the Taguchi method, to optimize its mechanical properties. Existing research predominantly relies on conventional methods, which may not fully capture the complexities involved in material optimization. This gap highlights the need for a more systematic approach to understanding how different processing parameters influence the mechanical properties of PKS composites.

Fiber reinforcement composites are particularly valued for being nonconductive, corrosion-resistant, impact-resistant, and lightweight [6]. Common sources of fibrous agricultural waste include rice husk, bagasse, oil palm fiber, and kernel shell, which are used in eco-friendly natural fiber composites due to their renewable supply, biodegradability, low density, low energy consumption, and abrasion resistance [7]. Popular thermosetting resins among scientists, researchers, and manufacturers include epoxies, polyester, and phenolics. Orthophthalic unsaturated polyester (UPE) stands out for its superior mechanical and thermal properties, room-temperature curing capabilities, cost-effectiveness, and [8]. The incorporation of palm kernel shell has been shown to enhance the tensile strength, elongation at break, and impact strength of polypropylene (PP) composites while reducing their tensile modulus.

Mechanical testing is crucial in the design and production of composites. These tests are necessary for product design and analysis, quality control, application performance requirements, and final product manufacturing. The tensile test is one of the most commonly used mechanical tests, as it assesses strength, ductility, modulus of elasticity, and strain hardening. Composite tensile testing adheres to standards such as ISO 527-4 and ASTM D638-14, which cover basic or flat sandwich tension. This testing measures various parameters, including modulus of elasticity, strain, elongation, and percent elongation at yield and break. The tensile test speed can influence overall tensile test results, with faster tests typically showing reduced tensile strength and other properties because the material has less time to deform and rearrange under stress. Conversely, slower testing speeds may lead to time-dependent deformation, affecting strength results. Thus, choosing a tensile test speed that matches the material and desired applications is essential for accurate findings. The driven grip or fixture's idle motion rate may be used when the testing speed is within the allowable variance, as referenced in ASTM D638-14.

23.3 Methodology

See Fig. 23.1 and Table 23.1.

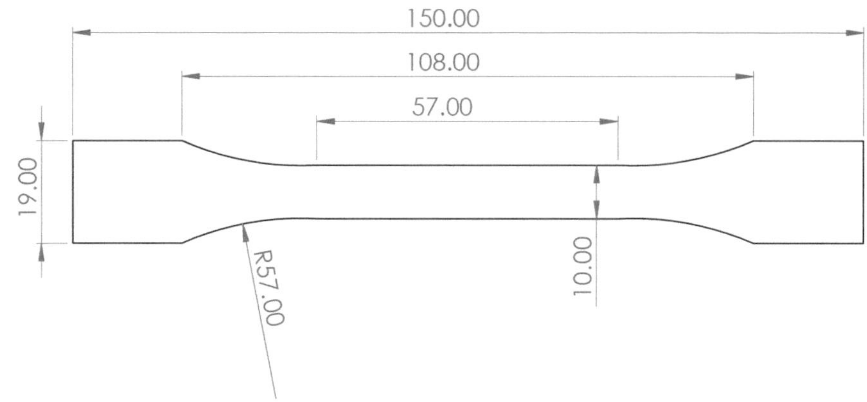

Fig. 23.1 The schematic sketch of the specimen

Table 23.1 Palm kernel shell composite composition

Number of specimens	PKS Particle (g)	Epoxy Resin (g)	Hardener (g)	Total (g)
1	3	5.25	1.75	10
2	3	5.25	1.75	10
3	3	5.25	1.75	10
4	4	4.50	1.50	10
5	4	4.50	1.50	10
6	4	4.50	1.50	10
7	5	3.75	1.25	10
8	5	3.75	1.25	10
9	5	3.75	1.25	10

23.3.1 Experimental Setup

The experimental setup is essential to the success of the experiment. Firstly, equipment, materials, and machinery must be prepared. Minitab was used for data analysis, and Solidworks was used to design the palm kernel shell composite specimen patent before creating the mold for specimen duplication. The raw palm kernel shell, epoxy resin, hardener, micro-weighing scale, and syringe for the tensile specimen must also be prepared. This experiment will test the specimens using a universal tensile testing machine. Table 23.1 shows the PKS composite composition.

Then, the SolidWorks software was used to redesign a square patent with a length of 190 mm, width of 160 mm, and thickness of 14 mm, as shown in Fig. 23.1. The designed data is then converted to STL file format suitable to the 3D printing machine requirement. Based on the specimen patent printed from the 3D printer machine, a mold is developed using the liquid silicon rubber mixed with the hardener to enable the exact duplication of specimens. Silicon rubber molds are used in this method due to their flexibility and robustness, non-stick and non-adhesive behavior, ensuring smooth specimen development.

Specimens 1, 2, and 3 require 30% palm kernel shell particle and 70% epoxy resin. Additionally, specimens 4, 5, and 6 need 40% palm kernel shell particle and 60% epoxy resin. The remaining specimens 7, 8, and 9 require 50% palm kernel shell particle and 50% epoxy resin. A micro-weighing scales measured the palm kernel shell particles. Epoxy resin and hardener were poured according to specimen weight after weighing the palm kernel shell particle. For 5 min, the epoxy resin and hardener were gently stirred in a cup. Next, the silicon rubber mold was filled with a balanced palm kernel shell, epoxy resin, and hardener mixture. The composites consolidated and hardened after 24 h at room temperature (20–25 °C). After 24 h, the palm kernel composites were carefully removed from the silicon rubber mold.

23.3.2 Mechanical Testing

The tensile test specimens, manufactured to adhere to the ASTM D638-14 standard, underwent mechanical testing. Specifically, the palm kernel shell composite specimens were subjected to mechanical testing to determine their modulus of elasticity, tensile strength, and elongation at break. These tests were conducted using the Galdabini Quasar 100 universal testing machine at the Universiti Kuala Lumpur Malaysian Spanish Institute. The Quasar 100 machine features a frame capable of withstanding a maximum load of 100 kN (22,481 lbs.). In this study, nine specimens were examined at different speeds of 5 mm/min, 6 mm/min, and 7 mm/min. To ensure accurate measurements, an extensometer was attached to the specimen using tensile test grips. It is worth noting that the ASTM method permits speeds ranging from 5 to 500 mm/min, depending on the specimen's shape.

23.3.3 Data Collection

Many experiments in various fields utilize the design of experiment (DOE) method to systematically collect and analyze data. In this study, a total of nine specimens were analyzed using the Taguchi method L9, which is a widely used approach to investigate different parameters while minimizing the number of experiments required. The Taguchi optimization technique was then employed to determine the optimal parameters of a composite specimen made of Palm kernel shell, based on the results obtained from tensile tests.

The analysis of the data obtained from the Taguchi method revealed the key factors influencing the tensile strength, elastic modulus, and elongation at the break of the composite specimen. The Taguchi method, developed by the renowned Japanese quality management expert Genichi Taguchi, is known for its efficiency in identifying significant factors in experimental designs. To ensure the reliability of the findings and identify potential sources of error or variation, a confirmation test, also referred to as a verification or validation test, was conducted. This statistical experiment, derived from the Taguchi technique, aims to validate the results obtained from previous research.

In order to determine if there were any significant differences among the means of different factor levels, analysis of variance (ANOVA) was employed. ANOVA is a statistical method commonly used to compare group means and assess the significance of differences observed. It can be applied to test theories related to product or process performance. In this study, the verification test involved comparing the signal-to-noise (S/N) ratios obtained from the optimal level design parameters with the projected ratios, expressed as a percentage.

23.4 Result and Discussion

A tensile test was conducted on composites made from palm kernel shells, employing the Galbidini Quasar 100 apparatus. The Taguchi method L9 was utilized to analyze a total of nine specimens, with an additional confirmation test. The application of Taguchi method L9 enabled the reduction of production time and the number of trials. In this experiment, the palm kernel shell specimen was subjected to tensile testing until failure, under different weight percentages and speeds. The resulting tensile strength, modulus of elasticity, and elongation at break were measured through the utilization of tensile testing. Table 23.2 shows the result of the testing.

Table 23.2 Results of tensile test using Taguchi orthogonal array L9

Experiment number	Speed (mm/min)	Parameter		Modulus of elasticity (N/mm^2)	Elongation %
		Weight percentage (%)	Tensile strength (N/mm^2)		
1	8	30	15.36	1178.47	0.55
2	9	30	17.59	1473.06	1.09
3	10	30	18.05	1329.62	3.14
4	8	40	16.95	2542.05	0.86
5	9	40	22.08	2445.23	1.04
6	10	40	25.06	2216.82	1.87
7	8	50	16.17	1782.89	0.46
8	9	50	8.12	1251.86	6.06
9	10	50	14.68	1383.74	1.79

23.4.1 Taguchi Data Analysis of Tensile Strength

The analysis of variance conducted to examine the relationship between tensile strength, weight percentage, and tensile test speed yielded non-significant results for both weight percentage ($p = 0.560$) and tensile test peed ($p = 0.159$). These findings indicate that neither weight percentage nor speed significantly impact tensile strength. Furthermore, the data failed to provide evidence to reject the null hypothesis that weight percentage does not affect speed. Consequently, this investigation concludes that weight percentages do not affect speed. Table 23.3 shows the analysis information.

Tensile mechanical characteristics were investigated in this study and compiled in Table 23.4. As shown in Fig. 23.2, the graph illustrates the effect of varying percentages (30%, 40%, and 50%) of PKS weight on the maximum tensile strength. The highest tensile strength of 25.06 N/mm^2 was observed at a 9 mm/min loading rate with 40% PKS weight. It is found that an increase in PKS weight percentage from 30 to 40% led to an increase in tensile strength, which can be attributed to the enhanced interfacial bonding between the matrix and PKS particles. Previous research conducted by Ramanaiah et al. [7] demonstrated that adding PKS particles to the matrix resulted in forming a rigid interface, thereby restricting matrix chain mobility and contributing to an increased composite tensile modulus. This phenomenon can be attributed to the higher stiffness of the composites. The strong connections between the hydrophilic PKS filler and the hydrophobic matrix facilitated stress transmission and improved material strength. Consequently, the composite with a higher PKS content exhibited a greater capacity to withstand applied loads without failure than the composite with a lower PKS content.

Table 23.3 Analysis of Variance (ANOVA) for SN tensile strength ratios

Source	DF	Seq SS	Adj SS	Adj MS	F	*P*
Weight Percentage %	2	7.088	7.088	3.544	0.67	0.560
Speed (mm/min)	2	31.794	31.794	15.897	3.02	0.159
Residual Error	4	21.038	21.038	5.059		
Total	8	59.920				

Table 23.4 Correlation of mean and standard deviation

Percentage	8 mm/min	9 mm/min	10 mm/min	Mean	Std Dev
30%	15.36	17.59	18.05	51	1.43878421
40%	16.95	22.08	25.06	64.09	4.10222297
50%	16.17	8.12	14.68	38.97	4.28283784

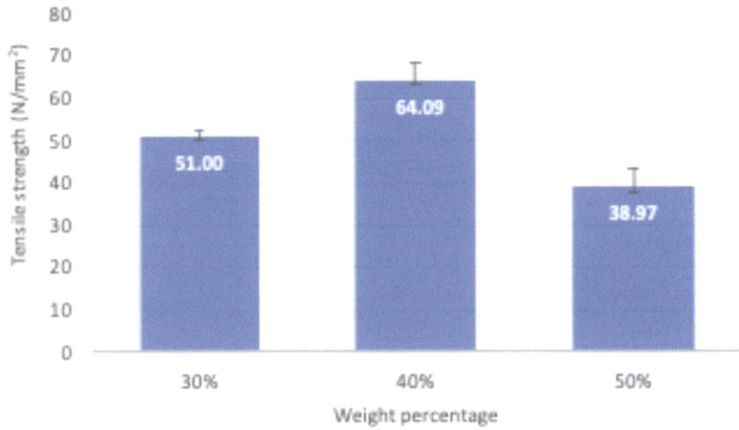

Fig. 23.2 The tensile strength versus weight percentage graph

23.4.2 *Taguchi Data Analysis of Modulus of Elasticity*

The modulus of elasticity quantifies the material stiffness. Stiff materials deform less during loading. From the tensile test data, modulus elasticity vs weight percentage and tensile test speed on ANOVA for SN ratios had *p*-values of 0.016 and 0.771, respectively. Tensile test speed is not significant, but modulus of elasticity is a significant characteristic. The modulus of elasticity is unaffected by tensile test speed. Data analysis showed that weight percentage and tensile test speed are significantly related. Thus, weight percentage affects the tensile test speed in this investigation. Table 23.5 shows the analysis information.

Figure 23.3 show the highest mean of modulus elasticity at 40% weight and 8 mm/min. Figure 23.3 also demonstrates that 40% had a higher modulus of elasticity than 30% but declined as weight percentage content increased to 50%. The matrix became stiff as PKS was added. 40% PKS lowers matrix molecular chain mobility. According to Sahari and Maleque, it will hinder matrix chain mobility and raise PKS tensile modulus. Composites are rigid and stiffer [9]. PKS at 50% of the composites had a lower modulus of elasticity than PKS at 40%, which increased the elongation at

Table 23.5 Analysis of variance for SN ratios of modulus elasticity

Source	DF	Seq SS	Adj SS	Adj MS	F	*P*
Weight percentage %	2	46.2209	46.2209	23.1104	13.77	0.016
Speed (mm/min)	2	0.9343	0.9343	0.4672	0.28	0.771
Residual error	4	6.7110	6.7110	1.6778		
Total	8	53.8862				

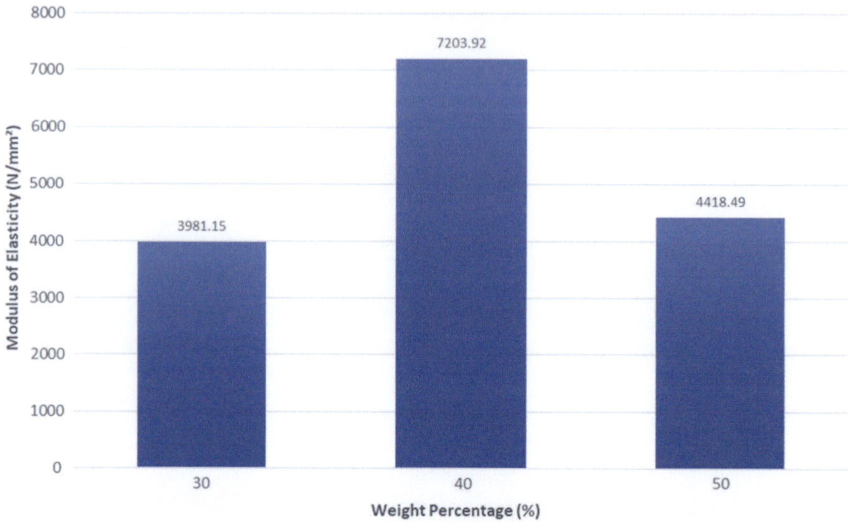

Fig. 23.3 The modulus elasticity versus weight percentage graph

break as a stiff material, decreasing the deformability of a rigid interface between the PKS and matrix and slightly changing its shape under elastic stresses.

23.5 Conclusions

The optimal values obtained at a weight composition of 40% and a speed of 10 mm/min were 22.06 N/mm^2 for tensile strength, 2216.82 N/mm^2 for modulus of elasticity, and 1.87% for elongation at break. The Taguchi technique can analyze the optimal parameters of PKS composite specimens with mechanical properties. Weight percentages, tensile test speed, and elongation are not considered significant parameters, as indicated by a p-value greater than 0.05. However, they are still utilized for validation purposes. The modulus of elasticity is considered significant due to the *p*-value being less than 0.05. The Taguchi investigation revealed that variations in weight percentage and tensile test speed have no significant impact on the tensile strength and elongation of PKS composites. The Taguchi investigation provides evidence for the alternative hypothesis that both weight percentage and tensile test speed significantly impact the modulus of elasticity.

References

1. Adeniyi AG, Onifade DV, Ighalo JO, Adeoye AS (2019) A review of coir fiber reinforced polymer composites. Compos B Eng 176:107305
2. Anyaoha KE, Mouazen AM, Sakrabani R, Patchigolla K (2018) Evaluating oil palm fresh fruit bunch processing in Nigeria. J Inst Agric Eng 36(3):236–246
3. Ganeshan R, Kulkarni S, Boone T (2001) Production economics and process quality: A Taguchi perspective. Int J Prod Econ 71(1–3):343–350
4. Atiqah A, Maleque MA, Jawaid M, Iqbal M (2014) Development of kenaf-glass reinforced unsaturated polyester hybrid composite for structural applications. Compos B Eng 56:68–73
5. Azreen Fuadi N, Saadi Ibrahim A, Nor Ismail K (2012) Review study for activated carbon from palm shell used for treatment of waste water. J Environ 1(5):252–266
6. Samotu IA, Dauda M, Anafi FO, Obada DO (2015) Suitability of recycled polyethylene/palm kernel shell-iron filings composite for automobile application. Tribol Ind 37(2):142–153
7. Ramanaiah K, Prasad AVR, Chandra KH (2012) Effect of fiber loading on mechanical properties of borassus seed shoot fiber reinforced polyester composites. Polym-Plast Technol Eng 3(3):374–378
8. Sahari J, Maleque MA (2016) Mechanical properties of oil palm shell composites. Int J Polym Sci

Chapter 24
Exemplified Integrity Assessment Using Phased Array Ultrasonic Testing on the Degraded 30" Rich Diglycolamine Lines at a Gas Processing Plant

Shahrizan Yusoff, Soleham Juraimi Kamaruddin, and Mohamad Sazali Said

Abstract The aim of this study is to establish an assessment method to assess the integrity of the degraded lines and to establish a mitigation or repair plan on deficiency of the 30" rich diglycolamine (RDGA) at gas processing plant. There are four lines of 30" diameter, transporting RDGA from the feed gas separation unit, gas sweetening and condensate stabilization, into the solvent regeneration. The scope of work of the candidate is limited to review the construction, testing and commissioning procedure for piping, operational history, performing evaluation of defects by the phased array ultrasonic testing, under consideration of operating conditions and evaluation of the repair method used on the RDGA line in terms of material selection, and perform fitness for service assessment. Based on this assessment study and proposed repair methods, these RDGA lines are fit for service to operate within their operating parameters and permanent repair should be in place by 2024.

Keywords Stress corrosion · Chloride · Petrochemical industry · Phased array ultrasonic testing

S. Yusoff (✉) · S. J. Kamaruddin · M. S. Said
Kulim Hi-Tech Park, Universiti Kuala Lumpur Malaysian Spanish Institute, 09000 Kulim, Kedah, Malaysia
e-mail: mshahrizan@unikl.edu.my

S. J. Kamaruddin
e-mail: soleham@s.unikl.edu.my

M. S. Said
e-mail: msazali@unikl.edu.my

T. A. A. Razak et al. (eds.), *Progress in Engineering Technology VI*, Advanced Structured Materials 229, https://doi.org/10.1007/978-3-031-81517-1_24

24.1 Introduction

Some literature said that maintenance is a necessary evil [1]. This is because maintenance is involving a huge number of money. However, due to the aging of equipment, addition and modification of technology, or the expansion of production units, the future of competitiveness for any petrochemical company is becoming more difficult. Maintenance costs can absorb 2–10 percent of a company's revenue and can reach up to 24 percent in the petrochemical industry [2]. In larger petrochemical refinery plants, lowering maintenance costs by $10 million adds as much to profitability as raising revenues by $30 million a year [3].

A pipeline is one of the important components of a gas processing plant. Petroleum pipelines transport crude oil or natural gas liquids, and there are three main types of petroleum pipelines involved in this process which are gathering systems, crude oil pipeline systems, and refined products pipelines systems [4]. The gathering pipeline systems gather the crude oil or natural gas liquid from the production wells. It is then transported with the crude oil pipeline system to a refinery. Once the petroleum is refined into products such as gasoline or kerosene, it is transported via the refined products pipeline systems to storage or distribution stations [5].

Only careful design and construction, combined with sufficient and routine maintenance, will ensure a structure's long and stable service life [6]. However, corrosion has always been a problem in the petroleum refining and the petrochemical operations [7]. The petrochemical process elements, such as furnace tubes, valves, and pipelines, frequently perform in chloride-containing, high-temperature environments, pitting, crevice corrosion, and stress corrosion cracking (SCC) are often associated with the operation [8]. In addition, it was found that the factors most affecting corrosion of structural materials in the petrochemical industry are chloride (Cl−) and hydrogen sulfide (H_2S) [9]. H_2S is an important constituent of refinery sour waters and is formed by the decomposition of organic sulfur compounds that are present at elevated temperatures (Fig. 24.1).

In this article, the challenge is to use and apply the phased array ultrasonic testing (PAUT) to examine the operating conditions and to evaluate the repair methods used on the RDGA line in terms of material selection. Then perform a fitness for service

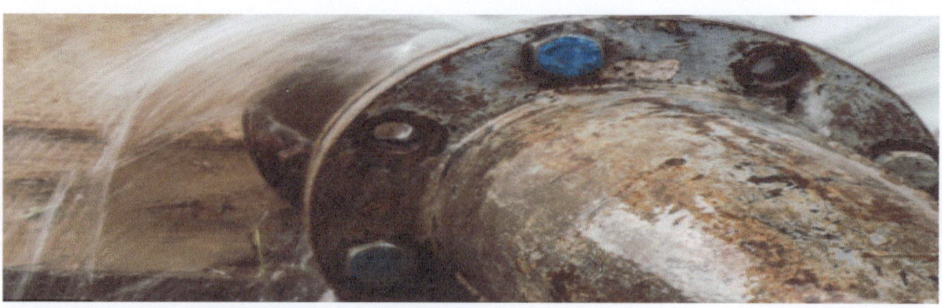

Fig. 24.1 Motor operate valve structure drawing

assessment in order to determine what, why, where, and how to repair. The model built in this study will assist a petrochemical company in implementing an efficient and optimal PAUT assessment strategy. The outcomes of the implementation process can be used to continually schedule development actions and provide better maintenance in the future. As a result, the organization will accomplish its mission and vision by implementing an overall optimal maintenance scheme.

24.2 Literature Review

The petroleum pipeline has been the subject of a variety of guidelines such as Petroleum (Safety Measures) ACT A807, which sets minimum design, management, and monitoring requirements for storage tanks in Malaysia [10]. In addition to general legislation, these facilities must meet special spill prevention, control, and countermeasure (SPCC) regulations. RDGA failure poses a major threat to the environment because it can pollute air and water and endanger human lives if it leaks or ruptures.

The aging management of key components, according to the American Petro Industry (API), is a major issue because some of the components have become obsolete and are no longer made, as well as the stringent regulatory process. According to the API STD 571, damage mechanisms are affecting fixed equipment in the refining industry. Below are critical factors for chloride stress corrosion cracking.

i. Chloride content, pH, temperature, stress, presence of oxygen, and alloy composition are critical factors.
ii. Increasing temperatures increase the susceptibility to cracking
iii. Increasing levels of chloride increase the likelihood of cracking.
iv. SCC usually occurs at pH values above 2. At lower pH values, uniform corrosion generally predominates. SCC tendency decreases toward the alkaline pH region.
v. Cracking usually occurs at metal temperatures above 140 °F (60 °C), although exceptions can be found at lower temperatures.
vi. Stress may be applied or residual. Highly stressed or cold-worked components, such as expansion bellows, are highly susceptible to cracking.
vii. Oxygen dissolved in the water normally accelerates SCC but it is not clear whether there is an oxygen concentration threshold below which chloride SCC is impossible.
viii. Nickel content of the alloy has a major effect on resistance. The greatest susceptibility is at a nickel content of 8–12%. Alloys with nickel contents above 35% are highly resistant and alloys above 45% are nearly immune.

Chloride stress corrosion cracking can be identified by its appearance and morphology described below:

i. Surface breaking cracks can occur from the process side or externally under insulation.

ii. The material usually shows no visible signs of corrosion.
iii. Characteristic stress corrosion cracks have many branches and may be visually detectable by a craze-cracked appearance of the surface.
iv. Fracture surfaces often have a brittle appearance.

24.3 Methodology

Refinery 514 was inspected in accordance with API STD 571 (2014 Edition + Addendum 1: 2018) for one of the petrochemical companies. In this study, the following factors are considered significant contributors in creating and maintaining the company's competitiveness:

i. High-quality product/service;
ii. Delivery on time;
iii. Environmental friendly production;
iv. Society acceptance.

In order to find the best combination of PAUT activities each time, an RDGA pipeline has also been the subject of a variety of guidelines such as Petroleum (Safety Measures) ACT A807, which is a framework that was proposed to provide an integrated safety measures system at the petrochemical refinery facility. Carefully designed operations will eliminate job waste, while proper maintenance preparation will ensure that the organization is adequately resourced in terms of human resources and replacement parts for the activities. Figure 24.2 shows the proposed framework used for this research.

This analysis is to focus inspection activities on those pieces of equipment when failure risks associated with an active damage mechanism are highest. The analysis is performed through the following steps.

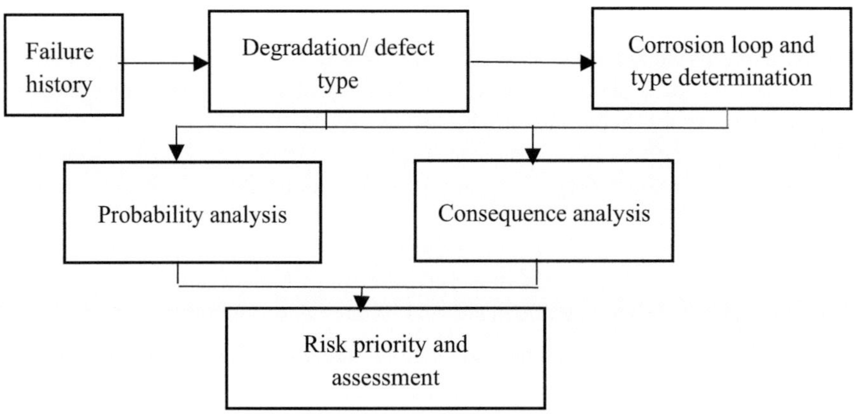

Fig. 24.2 The framework used in this research

i. Identification of the main active damage mechanisms based on failure history;
ii. Evaluation of defect using PAUT method and assessment of crack-like flaws for each piece of equipment as a function of the different damage mechanisms it is exposed to;
iii. Calculation of the level 1 fitness for service (FFS) for consequence of failure associated with each piece of equipment;
iv. Combination of the probability and the consequence numbers to calculate the maximum permissible crack lengths and evaluate results.

24.4 Result and Discussion

In summary, by identifying the primary active damage mechanisms based on the failure history, the maintenance teams had successfully identified the issue. The visual inspection of the initial leak revealed that it occurred at the weld joint of the pipe, allowing the leak to reach the pipe surface and release into the atmosphere. Clear evidence of this leak is observable as RDGA continuously flows from the internal pipe, as illustrated in Figs. 24.3 and 24.4.

Fig. 24.3 Defect on the structure

Fig. 24.4 RDGA flowing

Based on the findings from first leak, it was suspected that the chloride stress corrosion cracking (Cl-SCC) is the damage mechanism that attacked the RDGA lines. It can be observed by surface-initiated cracks on the weld joint caused by environmental cracking of 316 L series SS under the combined action of tensile stress, temperature, and an aqueous chloride environment. Impurities such as chloride gradually build up in DGA systems until a steady-state concentration is reached. The presence of dissolved oxygen and residual stress on the weld increases propensity for pitting and cracking.

Then, the next step is to evaluate the defect. There are 1711 welded joints in total on all four RDGA lines and all of them were inspected by using the PAUT method during construction phase. Also, all failures have occurred on a section between battery limits of units 731–732 to the section adjacent junction between units 780 and 753. 111 joints have been inspected with PAUT on both lines. Based on the PAUT inspection, it was observed that 21 and 9 joints have crack-like flaws on line 0001 and line 0002 respectively. Hence, the final inspected joints and results are tabulated in Table 24.1.

Based on the PAUT results, a level 1 fitness for service (FFS) assessment was completed for those 30 joints as per API 579-1/ASME FFS-1 2007, Part 9 Assessment of Crack-Like Flaws. This compares the relative flaw tolerance of an existing component for screening purposes. Using the results of the level 1 FFS, it was then determined the number of weld joints potentially with future risk of failure/leakage.

The presence of the aforementioned three joints found within a sample of 111 joints results in a relative frequency of 0.027, or 2.7%. There are 111 joints within the sections that have experienced leaks previously, so it could be inferred that among two lines in total, there is one joint that will have flaws exceeding the screening criteria. Because already 13 failures (leaks) have occurred, and it can be concluded that the probability of a leak happening at any of the two lines is very low since the majority of the susceptible joints have already failed.

Finally, investigation revealed concerns during the construction stage with the quality of welding such as a high repair rate, discrepancies on the two welding procedure specifications (WPS) used, welded joints found with rusting, radiographs with poor quality and not able to identify defects, concerns on pickling and passivation, among other findings. The most probable cause is a failure to adequately drain the water used for hydro testing of these lines, during the construction phase. This water, despite having records showing it met specifications, could not be adequately drained resulting in sections with stagnant water which progressively evaporated in the span of the seven months prior operation. The chloride concentration in this water would

Table 24.1 Tabulation of the inspected joints

Line number	Inspected joints	Crack like flaws	Leaked joints
0777-0001-RSX-30"	69	21	7
0777-0002-RSX-30"	42	9	6
Total	111	30	13

have increased significantly until it initiated pitting on the 316 L material. Flawed welded joints would have presented areas of less resistance to pitting and were much more affected.

24.5 Conclusion

Despite the RDGA pipeline oscillation problem, the data from the PAUT assessment on the RDGA surrounding the reactor are still safe according to the API STD 571 guideline. As a result, no further investigation is needed at this time. The principle that the fundamental positive-sequence active power absorbed by the RDGA can be used to monitor the mechanical state of this device is also based on this approach. FFS for PAUT is rather motivating and allows for monitoring the mechanical state of one specific part of the RDGA. For future works, on the other hand, the permanent repair consists of the removal of all existing composite wraps and mechanical clamps on all lines and repair of joints by grinding out defective welds and re-welding as per the approved procedure. These activities include reducing the process load and recirculating the RDGA to minimize H_2S content, hence reducing risk in subsequent steps. The joints to be repaired are located in a straight section located between two expansion loops that are roughly one meter higher than the straight section, facilitating drainage.

In conclusion, this system effectively guarantees a continuous secure operation of the facilities and can prevent catastrophic failure and loss of containment, according to this PAUT evaluation. Installing a wireless condition monitoring sensor for online monitoring and installing viscous dampener support at the possible position on the RDGA to minimize vibration are two changes that can be recommended in this report.

References

1. Al-Najjar B (2007) The lack of maintenance and not maintenance which costs: a model to describe and quantify the impact of vibration-based maintenance on company's business. Int J Pro Eco 107(1):260–273
2. Butkovic V, Cerjak J (1999) Implementation of motor operated valves (MOV) program in NE Krsko 453–460
3. Chang MK, Chang RR, Shu CM, Lin KN (2005) Application of risk based inspection in refinery and processing piping. J Loss Prev Process Ind 18(4–6):397–402. https://doi.org/10.1016/j.jlp.2005.06.036
4. Cholasuke C, Bhardwa R, Antony J (2004) The status of maintenance management in UK manufacturing organisations: results from a pilot survey. J Qual Maint Eng 10(1):5–15
5. Department of Occupational Safety and Health Laws of Malaysia Act 302 Petroleum (Safety Measures) Act 1984 (2006)

6. Hale S (2003) Improvements in valve reliability due to implementation of effective condition monitoring programs. The Proc Int Conf Nuclear Eng (ICONE) 2003:387. https://doi.org/10.1299/jsmeicone.2003.387
7. Kim DW, Park SG, Kang SC, Lee SG, Hong SY (2011) Analysis of the effect of lubrication performance on motor operated valve actuator output thrust. Nucl Eng Des 241(8):2716–2721. https://doi.org/10.1016/j.nucengdes.2011.06.007
8. Krishnasamy L, Khan F, Haddara M (2005) Development of a risk-based maintenance (RBM) strategy for a power-generating plant. J Loss Prev in Process Ind 18(2):69–81. https://doi.org/10.1016/j.jlp.2005.01.002
9. Mohamed RB, Hassan CRBC, Hamid MD (2018) Developing a risk-based inspection practices maturity model for Malaysian industries. Loss Prev Process Ind J 56:217–230. https://doi.org/10.1016/j.jlp.2018.08.011
10. Pewkliang S, Janbanjong P, Rahong P, Muangsuankwan N, Nokjib S (2017) The study of effects from small bore connection geometry for high vibrational excitation. In: Society of petroleum engineers—SPE Abu Dhabi International Petroleum Exhibition and Conference 2017. https://doi.org/10.2118/188452-ms

Chapter 25
Risk-Based Inspection Methodologies for Oil Refineries

Shahrizan Yusoff, Soleham Juraimi Kamaruddin, and Mohamad Sazali Said

Abstract This study is related to the risk-based inspection (RBI) of pressure relieving devices (PRD) for Malaysian petrochemical plant. The PRD RBI study was executed utilizing the API RBI software, which follows the API 581: 2008, Risk-Based Inspection Technology standard. API 581 specifies the quantitative approach in determining the risk for PRD. PRD technical datasheet, inspection reports, process flow diagram, protecting equipment technical datasheet, piping and instrumentation diagram, and mass and energy balance are among the essential documents required to perform this analysis. A total number of 169 PRD were analyzed. The PRD testing interval at an oil refinery is currently between 1 and 3 years. From the study, 72% (122 units) of PRD interval can be extended from existing interval practiced by PP(T)SB. Another 18% (31 units) of PRD interval needs to be reduced from existing interval practice and 10% (16 units) of PRD interval is unaffected. Several significant findings were noted throughout this study such as set pressure higher than equipment design pressure and data discrepancy on technical documents. PP(T)SB is strongly recommended to take all the necessary actions to ensure PRD is operating as intended.

Keywords Pressure relieving devices · Risk-based inspection · API 581 · API RBI software

---

S. Yusoff (✉) · S. J. Kamaruddin · M. S. Said
Kulim Hi-Tech Park, Universiti Kuala Lumpur Malaysian Spanish Institute, 09000 Kulim, Kedah, Malaysia
e-mail: mshahrizan@unikl.edu.my

S. J. Kamaruddin
e-mail: soleham@s.unikl.edu.my

M. S. Said
e-mail: msazali@unikl.edu.my

T. A. A. Razak et al. (eds.), *Progress in Engineering Technology VI*, Advanced Structured Materials 229, https://doi.org/10.1007/978-3-031-81517-1_25

25.1 Introduction

Petrochemical facilities and refineries are made up of hundreds of pressure vessels (shown in Fig. 25.1), heat exchangers, towers, and other units that work in a variety of ways. Corrosion, aging, wear, and other factors cause them to deteriorate. Furthermore, the established management criteria play a role in the process safety of a petrochemical process. It will have an impact on the plant's lifecycle, production activity, and eventual dismantling capability. The number of maintenance activities will be low at the time of failure if the consequence of failure is minimal. If the failure outcome is severe, however, equipment deterioration will result in unplanned shutdowns, productivity losses, high maintenance costs, and serious safety, health, and environmental concerns. If the degradation condition of plant equipment or pressure vessels can be modeled, the time to failure can be predicted, and appropriate inspection activities can be planned based on the service age and the expected failure time. A risk-based inspection not only increases the time between shutdowns but also saves millions of dollars.

Based on the research conducted by [1], 15% of the total production cost is spent on reliability research activities. On the other hand, [2] stated that reliability research including maintenance costs could reach 15–40% of production costs, varying according to the type of industry. Consequently, further research by [3] showed that up to 33% of this maintenance cost is actually wasted or spent unnecessarily. These percentages show many improvements could be carried out in achieving an effective and optimized maintenance system. Costly failures can be avoided by establishing systematic maintenance and risk-based inspection can be done properly.

This study expands on previous studies in order to understand the relationship between plant performance and asset effectiveness and reliability. The focus is to determine the optimum inspection interval for all PRD in scope, using the highly quantitative methodology in API 581 Year 2008 standard. To achieve this goal, this is designed to meet the following main objectives.

i. Review and understand the current inspection and maintenance plan.

Fig. 25.1 PRD asset

ii. Evaluate and agree on the appropriate quantitative risk target consistent with the qualitative approach.
iii. To provide new inspection intervals based on the risk target defined.

25.2 Literature Review

Most RBI researchers recently published an article on maintenance techniques to reduce the risk of failure. The term risk, according to [4], is a combination of the likelihood of failure and the consequences of failure that may be caused by a variety of causes, including material loss, device error, and natural hazards.

However, according to [5], successful asset management requires not only striking a balance between financial and nonfinancial metrics, but also covering two essential aspects of asset management. The first is primarily concerned with engineering aspects of information and communication technology, in which properties are handled by tracking their condition to ensure that they are in good working order.

25.2.1 Regulations, Codes, and Standards

All unfired pressure vessel (UPV), steam boiler, or hoisting machine other than a hoisting machine driven by manual power shall hold a valid certificate of fitness (CF) in respect thereof in Malaysia, based on jurisdictional requirements, such as the Factories and Machinery Act 1967 P.U. (A) 43/70, Factories and Machinery (Notification, Certification of Fitness and Inspection) Regulation, 1970. (Malaysia, Factories and Machinery (Notification, Certificate of Fitness and Inspection) Regulations, 1970. The validity of every certificate of fitness shall ordinarily be fifteen calendar months from the date of inspection or such longer period not exceeding three years as the chief inspection may consider appropriate in his discretion.

The federal government also regulates PRDs under ACT 302, Petroleum (Safety Measures) ACT A807, which establishes minimum design, management, and monitoring standards for storage tanks (Department of Occupational Safety and Health, 2006). These facilities must follow special spill prevention, control, and countermeasure (SPCC) regulations in addition to general regulations. Since furnace tubes can pollute lakes and rivers and danger human lives if they leak or rupture, they pose a serious threat to the ecosystem.

There have also been a number of international guidelines developed for furnace tubes. The American Petroleum Institute (API) established and published the "Tank Inspection, Repair, Alteration, and Reconstruction (API 653)" which provides instructions for the inspection, repair, alteration, and reconstruction of furnace tube used in the petroleum and chemical industries [6]. While the Steel Tank Institute/Steel Plate Fabricators Association (STI/SPFA), a trade association that represents steel construction product fabricators, released the STI SP001 in 2000, which covers

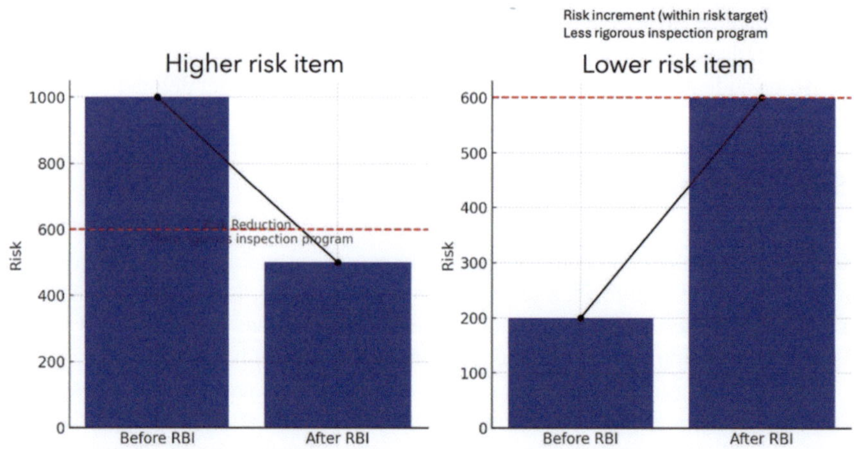

Fig. 25.2 Risk before and after RBI

inspection standards for welded metal, shop-fabricated, and small field-erected tanks, among other tubes.

25.2.2 Risk-Based Inspection Calculator

The main driver in any risk-based inspection (RBI) initiative is managing the risk within the acceptable limit. An inspection plan is then developed to reduce the risk of items which exceed the acceptable risk target [7]. For items whose risks are inherently low at the current inspection program, the inspection can be adjusted and typically could have been prolonged in order to optimize the cost and resources while maintaining them within the risk target, as shown in Fig. 25.2.

The overall risk for PRD is expressed in a quantitative value, either in area (m^2) or financial ($) based on two modes of failure; leakage during service and fail-to-open on demand for all overpressure scenarios [8]. Moreover, the leakage risk is a function of the probability of leak (POL) and consequence of leak (COL), and is always expressed in financial ($).

$$\text{RiskPRD} = \text{RiskLeak} + \sum_{1}^{n} \text{RiskFTO} \tag{25.1}$$

$$\text{COL} = \text{Cost}_{\text{inv}} + \text{Cost}_{\text{env}} + \text{Cost}_{\text{sd}} + \text{Cost}_{\text{prod}} \tag{25.2}$$

The COL is the weighted average between mild (small) leak and stuck-open (big) leak.

Cost_{inv} Cost of loss of inventory during leakage prior to being discovered.

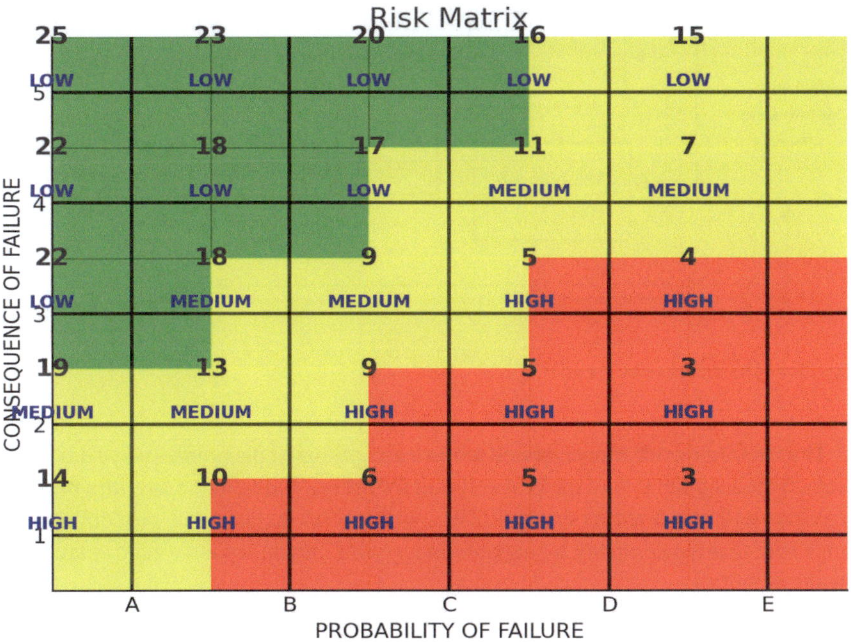

Fig. 25.3 RBI risk matrix

$Cost_{env}$	Environmental clean-up cost.
$Cost_{sd}$	Cost to repair a leaking PRD during shutdown.
$Cost_{prod}$	Cost of loss of production to repair the leaking PRD.

On the other hand, the results of the RBI analyses can be conveniently presented in a five-by-five risk matrix with the likelihood category of 1 to 5 and the consequence category of A to E [9]. The breakdowns of the probability and consequence categories are as given in Table 1 while the graphical representation of a 5 × 5 risk matrix is shown in Fig. 25.3.

25.3 Methodology

In this study, the following factors are considered significant contributors in creating and maintaining the company's competitiveness.

i. High-quality product/service.
ii. Delivery on time.
iii. Environmentally friendly production.
iv. Society acceptance.

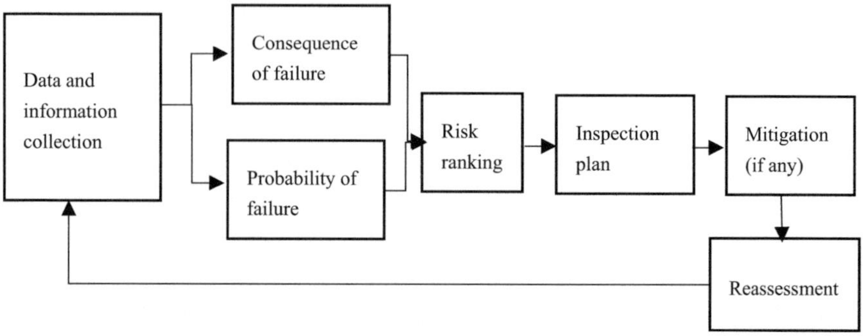

Fig. 25.4 The model assessment

Thus, a framework was proposed to have an optimized maintenance system at the petrochemical plant to find the best solution for prevention failure. Carefully planned activities will avoid redundancy of tasks, while good maintenance scheduling will make sure that the company is prepared in terms of human resources and spare parts for the activities.

Figure 25.4 shows the flow chart for the assessment. From the figure, there are six stages of assessment before corrective action is presented as a summary for improvement. The detailed results of the assessment will be discussed in the next section.

25.4 Result and Discussion

The current practice for oil refineries is between 1 and 3 years. Based on the analysis, 18% (31 units) of PRD are recommended to have an inspection interval between 1 and 1.5 years. Most of the PRD-protecting column systems in oil refineries were analyzed to be in this group of intervals. Another 20% (33 units) are recommended to be inspected and overhauled between 1.5 and 5 years and 13% (22 units) are recommended to be inspected and overhauled between 5 and 9.9 years, respectively. And the remaining 49% (83 units) of PRD can be tested at a maximum interval of 10 years. A summary of the results is presented in Fig. 25.5.

On the other hand, by implementing the newly calculated interval for PRD inspection, the cost–benefit for the oil refinery is estimated to be USD 59,476.46 (RM 180,600), which is 23% reduction from inspection cost over a 10 year period excluding the maintenance cost. This amount is obtained by comparing the frequency of overhaul and inspection of PRD before and after the RBI study. Default average overhaul cost per PRD was used to estimate the value creation from this study roughly. The value used for cost–benefit analysis is the same as the default value for inspection cost used in the financial consequence analysis for this study which is USD 1000 for

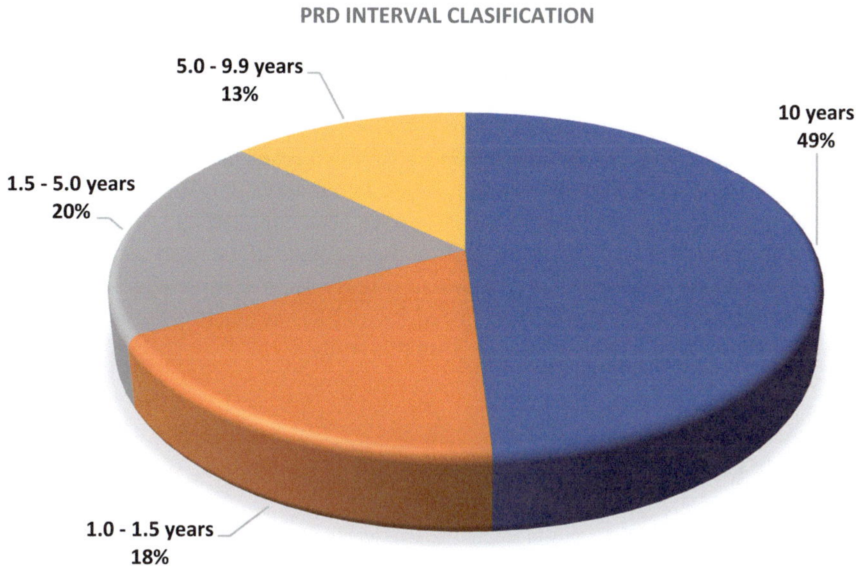

Fig. 25.5 Summary of PRD interval classification

less than 6-inch PRD and USD 2000 for more than 6-inch PRD. Figure 25.6 shows the comparison between two interval costs.

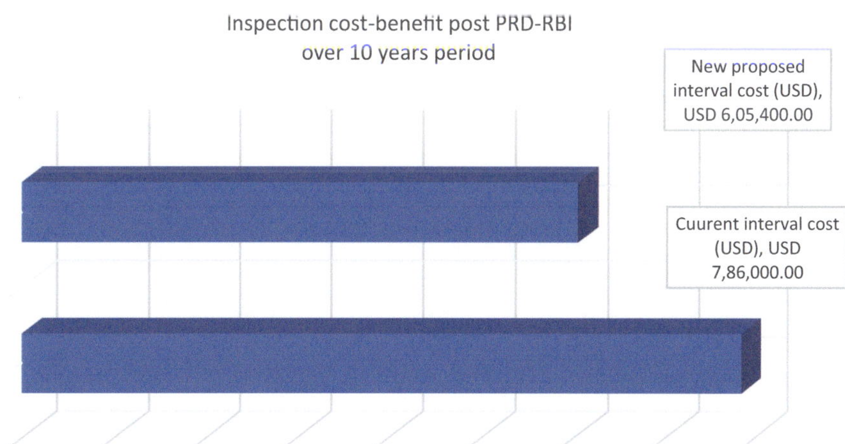

Fig. 25.6 Inspection cost–benefit

25.5 Conclusion

The data from the RBI assessment on the PRD shows that the damage state (or probability rating) of the protected equipment, PRD latest inspection results, and the overall process system are reviewed periodically. All relevant information consists of PRD design data and specification; protected equipment data, inspection history as well as applicable overpressure scenario have been captured and properly documented through this study. However, some recommendations such as to obtain a more accurate demand rate for overpressure scenario, a comprehensive capacity study/overpressure review is recommended for all PRD as this is essential to ensure that the PRD will gain the full benefit from this study.

References

1. Chang JI, Lin CC (2006) A study of storage tank accidents. J Loss Prev Process Ind 19(1):51–59
2. Chang MK, Chang RR, Shu CM, Lin KN (2005) Application of risk based inspection in refinery and processing piping. J Loss Prev Process Ind 18(4):397–402
3. Department of Occupational Safety and Health (2006) Laws of Malaysia Act 302 Petroleum (Safety Measures) Act 1984
4. Faber MH (2002) Risk-based inspection: an introduction. Struct Eng Int J Int Assoc Bridge Struct Eng (IABSE) 12(3):186–198
5. Fesenko H, Kharchenko V, Zaitseva E (2019) Evaluating reliability of a multi-fleet with a reserve drone fleet: an approach and basic model. Proceedings Int Conf Info Digital Technol 2019:128–132
6. Kabir S, Papadopoulos Y (2018) A review of applications of fuzzy sets to safety and reliability engineering. Int J Approximate Reasoning 100:29–55
7. Krishnasamy L, Khan F, Haddara M (2005) Development of a risk-based maintenance (RBM) strategy for a power- generating plant. J Loss Prev Process Ind 18(2):69–81
8. Mohamed RB, Hassan CRBC, Hamid MD (2018) Developing a risk- based inspection practices maturity model for Malaysian industries. J Loss Prev Process Ind 56:217–230
9. Moradi R, Groth KM (2019) Hydrogen storage and delivery: review of the state of the art technologies and risk and reliability analysis. Int J Hydrogen Energy 44(23):12254–12269

Chapter 26
Design and Development of a Roasted Groundnut Peeling Machine

W. Faradiana W. Maidin, Fazidah Saad, Zulkarnain Abdul Latiff, Noorhelinahani Abu Bakar, and Sharmiwati Mohammed Sharif

Abstract In this research work, an improved roasted groundnut skin peeling machine was successfully designed and developed. The main idea of the machines is to rotate the groundnut inside the drum peeler with the power of the rotary motor. The drum peeler wall has a shaft located in the middle of it. There are brushes attached to the rod, and these brushes will also be rotated to brush the groundnut skin and peel it off. The skin then will be separated from its nuts by the air blown in the drum and fall due to the gravitation effect. The use of this machine can reduce the cost of peeling the groundnut skin over the manual groundnut peeler. This machine is recommended for peeling the groundnut skin at small-size kitchens and at small-size enterprises only as there are a lot of machines that are focused on the industry level.

Keywords Groundnut peeling · Groundnut skin · Drum peeler

W. F. W. Maidin (✉) · F. Saad · N. A. Bakar · S. M. Sharif
Mechanical Section, Universiti Kuala Lumpur, Malaysian Spanish Institute Kulim Hi-Tech Park, 09000 Kulim, Kedah, Malaysia
e-mail: wfaradiana@unikl.edu.my

F. Saad
e-mail: fazidah@unikl.edu.my

N. A. Bakar
e-mail: noorhelinahani@unikl.edu.my

S. M. Sharif
e-mail: sharmiwati@unikl.edu.my

Z. A. Latiff
Manufacturing Section, Universiti Kuala Lumpur, Malaysian Spanish Institute Kulim Hi-Tech Park, 09000 Kulim, Kedah, Malaysia
e-mail: zulkarnain@unikl.edu.my

T. A. A. Razak et al. (eds.), *Progress in Engineering Technology VI*, Advanced Structured Materials 229, https://doi.org/10.1007/978-3-031-81517-1_26

26.1 Introduction

The seeds or the groundnut are a nutrient-dense food that is high in fat and protein. Groundnut is classified as both a grain legume and oil crop due to its high oil content. There are numerous usages of the groundnut and it can be classified as two main categories which are for peanut oil production and for food production. Besides groundnuts are used to make oil, and frequently used to make soups, sweets, roasted peanuts, confections, snack foods, and extenders for meat products. There are many different ways to consume groundnut around the world but most of which are traditional dishes. For this reason, the production of groundnut has increased globally. In 2020, there were roughly 47 million metric tons of groundnut produced worldwide, with China being the main producer [1]. Approximately 25–30% of the groundnut production is exported by the fourth-largest producer which is the United States and about 56% of the groundnut that were cultivated in 2020 were used to make peanut butter.

As mentioned earlier, the groundnut is used for oil and food production but actually, around 67% of the total groundnut production is used for making oil [2]. There are numerous studies [3–6] done on the design of groundnut processing machinery. All these ideas have been taken into consideration in this work as well in order to produce a better design to peel and separate the roasted groundnut skin.

The blanching method is one of the popular methods used to remove the groundnut skin. It uses the dry route consisting of sandpaper friction rollers which help to peel the peanut's skin and automatically separate the skin and peanuts with the help of a blower. In addition, various peanut peeling machines with different capabilities can be put into a complete processing row. As the Akintade-Bratte journal [5] reports, three main parameters are considered for the purpose of this study: blanch clearance, blanch speed, and feed rate. Blanching was performed with the use of a brush-like projection that was made from iron sheet, and the brush clearance was adjusted following the analysis results. Friction is applied with the rolled shape from the iron plate, and the roasted groundnut friction is rubbed continuously among the diameters in the inner drum.

The research work from Gana [7] uses the auger method to design, fabricate, and evaluate the performance of roasted groundnut skin peeling and also separating the skin. The process of the auger method is by running the roasted groundnut through the drum from the hopper to the inner drum that contains auger and brush that were attached. The roasted groundnut will feed between the inner and outer drum which keeps the groundnut rotating until the skin is peeled off. The separating and cleaning unit consists of an enclosure blower which works by blowing the skin out and groundnut that has been peeled will be separated. The highest skinning efficiency of this machine is 99.91% and was obtained when the machine was processed with six brushes and inner drum clearance of 1.4 cm while the lowest value of skinning efficiency of 7.9% was obtained when the sample was processed with two brushes and inner clearance of 1.8 cm.

The portable rotary peeler design is a simple design [2] that is focused on the portable dry groundnut husk peeling machine. The working principle of this machine is simpler than other machines. The machine consists of a rotating peeler which is covered in a vertical cylindrical coat filled with a soft rubber where the groundnut is fed into the cylinder shell through the inlet into the cylinder and the roller rotates the groundnut seeds against the cylindrical shell wall for peeling.

Although there are many groundnut shelling machines on the market, they are enormous, expensive, and unsuitable for residential use. They are appropriate for use in large industries only. The other concern in peeling the groundnut skin is the separation of the skin from the nuts itself. Therefore, the motivation of this study is to design and develop a roasted groundnut skin peeling machine that can separate the skin as well after the peeling process.

26.2 Description of the Issue in Groundnut Peeling

Generally, many Malaysian cuisines need to use the roasted groundnut without skin. To categorize the user, small-size cooks (cook for a small family) and large-size cooks (food vendors) are the two main groups who need to remove the groundnut skin. The difference between them is the amount of peeled roasted groundnut that needs to be used in their cookery. Therefore, in order to understand the procedure and issues the user encounters during the groundnut peeling process, two methods have been used in the field research which are observing the user peeling the groundnut skin and also interviewing them about the procedure used.

Based on the findings, the small-size cooks are women who cook mostly for their families, whereas large-size cooks are food vendors who have to cook enormous quantities of food. From the interview conducted with them, the small-size cooks only require about 300 g of groundnut to prepare a meal; however, the large-size cooks require about 2 kg of it.

Until now, most of the Malaysian cooker regardless of the small size or large size still use the manual method of skin peeling for the roasted groundnut. This is because the cost of a machine to peel the groundnut skin is very high as it is automated and in a huge size. This is only suitable for industrial purposes. From the observation, there are two steps taken to peel the groundnut skin which are the skin peeling process and the skin separation process.

- Method used to peel the groundnut skin.

The small-size cooks and large-size cooks use different procedures in peeling the groundnut skin.

i. Small-size cooks will manually rub the roasted groundnut using a cloth between palms to separate the skins.
ii. Large-size cooks will manually put in the roasted groundnut in a hollow drain basket and rub them continuously in it.

- Method used to separate the groundnut skin from the nuts.

All users follow the same laborious procedure, known as winnowing, in which air is blown into the peeled skin, causing it to automatically come apart, in order to remove the skin from the nuts. Usually, the electric fan is the source of the airflow.

In conclusion, the majority of respondents disclosed that, depending on the weight of the groundnut, peeling its skin takes anywhere from 5 to 20 min. As previously stated the procedure is an arduous manual operation that can occasionally be unhygienic because it involves rubbing the peanut with a cloth and winnowing it. In order to reduce time and simplify the procedure, a roasted groundnut skin peeling machine that integrates the peeling and separating of the skins has been designed.

26.3 Design Process of Groundnut Peeling Machine

Figure 26.1 shows the flowchart of the design process for the groundnut peeling machine. It started with objective identification via the objective tree method and function analysis method. Then the employment of specifications is studied by performance specifications of the machine's function. In order to prioritize and understand the relationship of each function, quality function deployment methods are used. Once all the objectives have been clarified, the idea brainstorming for each function is done by using a morphological chart which is used later in the idea sketching. Last but not least to finalize the design concept, a weighted objective method is used to validate the final design.

26.3.1 Objective Tree Method

Firstly, the objectives of the groundnut peeling machine are identified by using the objective tree method. Figure 26.2 shows the objective tree method used in this study. This method highlights the objectives which may be achieved by several different

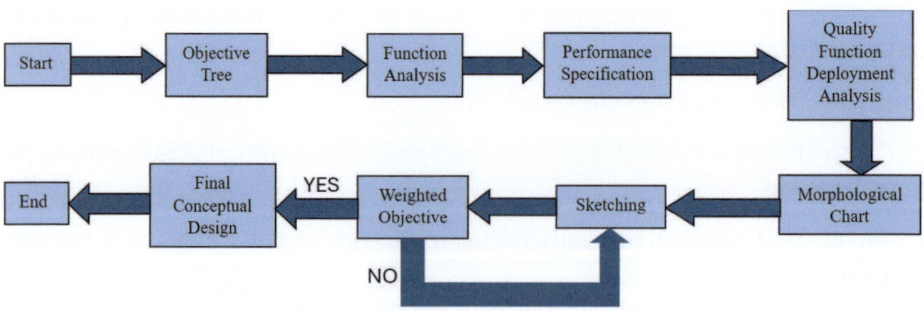

Fig. 26.1 Design process of groundnut peeling machine

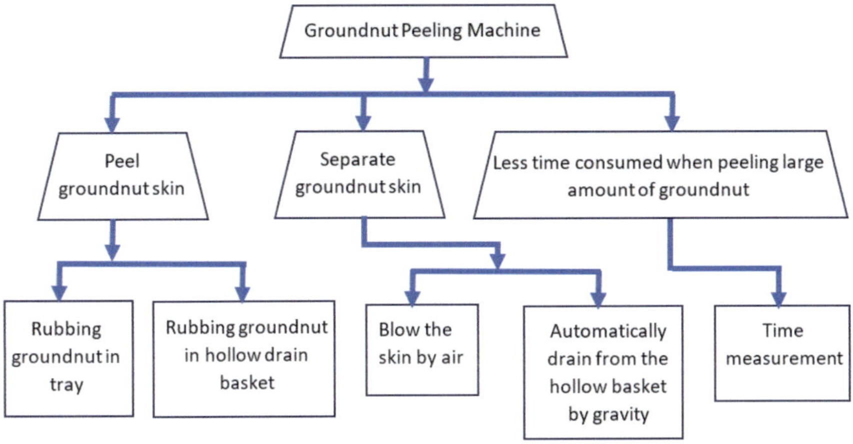

Fig. 26.2 Objective tree method

means or subobjectives. This application helps to ensure that all possible means and functions are considered in the design.

26.3.2 Function Analysis

The aim of the function analysis is to establish the functions required, and the system boundary of the new design is shown in Fig. 26.3.

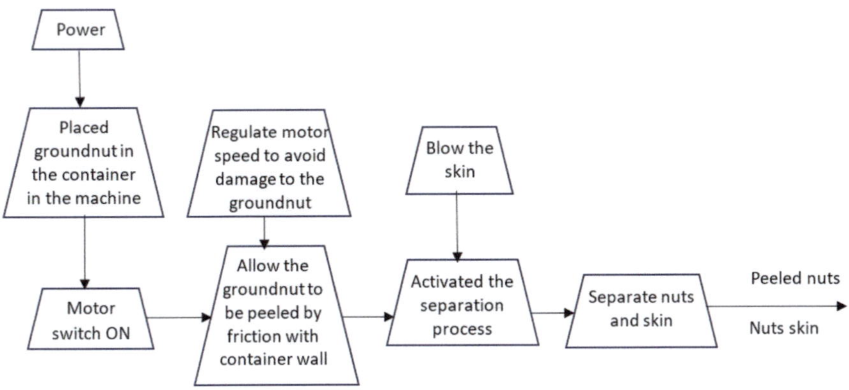

Fig. 26.3 Function analysis

26.3.3 Performance Specification

Performance specification is used to describe the specification of each task for the understanding the options in prioritizing the design constraint for this machine. This explanation shows the development of performance and lists the new product's attributes mainly in terms of a set of user needs as below.

1. Physiological needs:
   (a) Peel the groundnut skin.
   (b) Separate the skin and nuts.
2. Social needs:
   (a) Simple design that is easy to use and does not require complicated procedure.
3. Psychological needs:
   (a) The machine can do two processes which are peeling the peanut's skin and the separation.
4. Technical needs:
   (a) Lightweight.
   (b) Small scale.
   (c) Machine can be loaded at least with 2 ckg of peanuts.
5. Time needed:
   (a) Faster than manual operation.
   (b) Machine can be operated in less than 30 min.

26.3.4 Quality Function Deployment Analysis

The aim of the quality function deployment analysis is to set targets to be achieved for the engineering characteristics of a product, such that they will satisfy the design requirements. Since the design needs a machine to separate the skin from the nuts, therefore the design needs to have a dual function of separating the skin by providing a filter at the bottom of the machine instead of just focusing only to peel of the skin from the nuts. The result list of customer attributes and engineering characteristics, together with their relative importance rating could therefore be referred to as in Fig. 26.4.

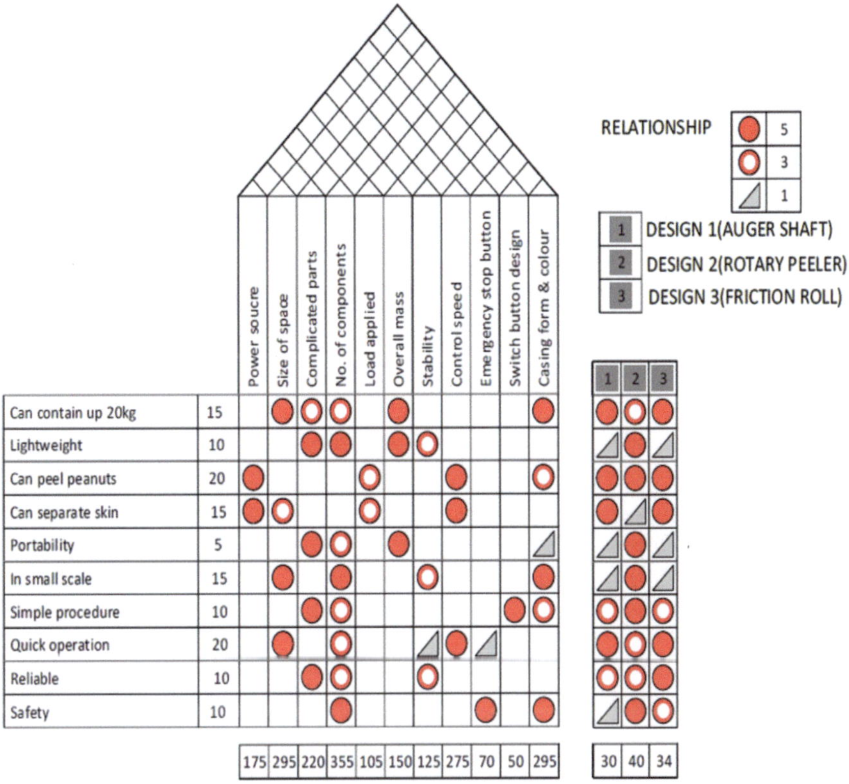

Fig. 26.4 Quality function deployment analysis

26.3.5 Morphological Chart

The morphological chart is used to identify the essential generic features for the peanuts peeling skin machine. The features are control of speed, capability of peeling peanuts, capability of separating the skin, weight of the machine, and ability to support load and components on small scale.

From the features listed above, Fig. 26.5 shows the morphological chart used for the design of the groundnut peeling machine. The circled means is chosen to proceed the design.

FEATURES	MEANS			
Power	Battery	Universal motor	-	-
Friction motion	Auger shaft	Rotary peeler	Friction roll	-
Space	Horizontal cylinder	Horizontal square	Vertical cylinder	Vertical square
Operator	Remote control	Button	-	-
Material body & casing motor	Stainless steel	Plastic	Glass	-
Material components	Stainless steel	Mild steel	Rubber	-

Fig. 26.5 Morphological chart

26.3.6 Ideas Sketching

Design 1 (Auger Shaft)

The first design is using the concept of the auger shaft that is just similar to the proposal by Gana [7] but it is in the small scale from the industry as the objective of this machine is to produce machines that are suitable for small-size cooks or small medium enterprise. Usually, this concept has been used in larger industries.

The working principle of this machine is that the roasted groundnut is fed to the machine where the inner drum contains the auger shaft and the brushes are attached. The brushes serve to remove the skin, and the outer drum is the enclosure for the inner drum which holds the groundnuts. Then, the function of the filter will be used to separate the skin and nuts automatically.

Design 2 (Rotary Peeler)

This design aims to produce a simple design that can peel and separate groundnuts and the peeling concept is inspired by Ikechukwu [2]. A concept of vertical cylinders was created to save space for the machine. The electric motor is fitted to the base and connected with the peeler drum to direct rotate in the peeling space while the peeling chamber does not move. The groundnuts will be peeled in between the peeling chamber and peeler drum, and this will automatically filter the skin.

Design 3 (Friction Roll)

The friction roll is widely used in the industry where the peanuts will rub between the plates and solid cylinder. Usually industrial machines use three or four components of solid cylinders to get an efficient result, but this design only uses one solid cylinder for design simplicity with less parts in the design. The groundnut skin that is separated during the rubbing process will be automatically separated by using a function filter. The filter concept is inspired by the observation from the user during their manual operation of separating the skins.

26.3.7 Weighted Objective Analysis

Weighted objective analysis produces a weight of each criterion using a specific comparison process based on the benchmark or initial data. It is done for each alternative's performance on each criterion.

Table 26.1 shows the weighted objective analysis of the design of groundnut skin peeling machine.

From this analysis, it is clearly shown that design 2, which uses the rotary peeler concept, is chosen to proceed as the final design concept for the machine as it has the highest value of requirement scores compared to the other two.

Table 26.1 Weighted objective analysis

		Requirement score		
Criteria	Weight	Design 1	Design 2	Design 3
Can contain up to 20 kg	15	5	3	5
Lightweight	10	1	5	1
Can peel peanuts	20	5	5	5
Can separate skin	15	5	1	5
Portability	5	1	5	1
In small scale	15	1	5	1
Simple procedure	10	3	5	3
Quick operation	20	5	3	5
Reliable	10	3	3	5
Safety	10	1	5	3
Total		30	40	34

26.4 Results and Discussion

The selected concept design that has been chosen based on the suitable criteria for peeling of the groundnut skin process. The final design of the roasted groundnut skin peeling machine is shown in Figs. 26.6, 26.7, 26.8 and 26.9. The peeler drum and peeler chamber will be rotated during the process, and the groundnut skin will be separated by the function in the rod at the middle of the peeler chamber. The rod attached to the peeler drum will also be rotated to rub the groundnut skin to make sure the peeling process succeeds.

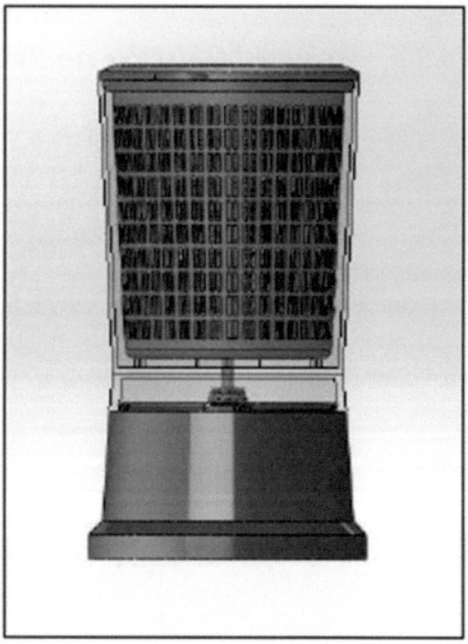

Fig. 26.6 Front view of the final conceptual design

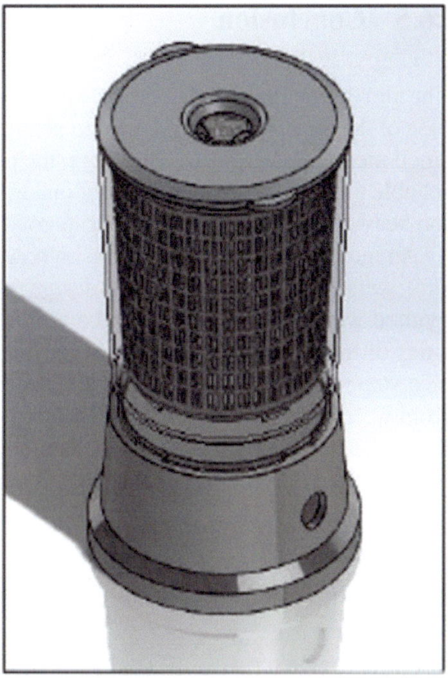

Fig. 26.7 Isometric view of the final conceptual design

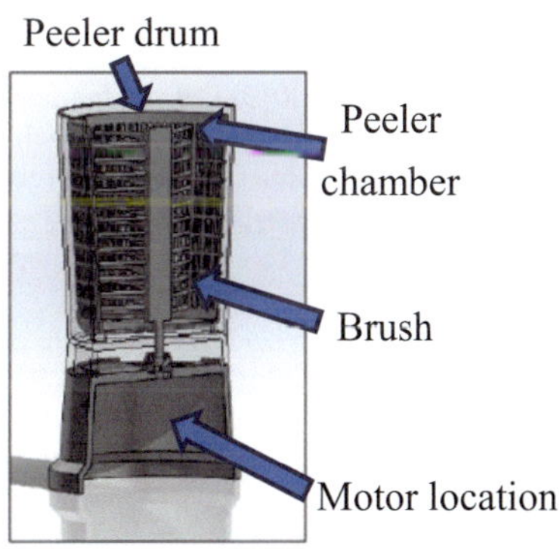

Fig. 26.8 Sectional view of the final conceptual design

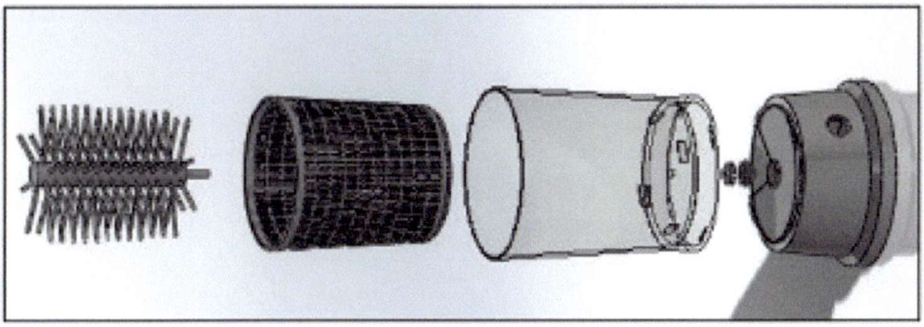

Fig. 26.9 Exploded view of the final conceptual design

26.5 Conclusion

The aim of this study was to design and develop a groundnut skin peeling machine to ease of the process of groundnut skin peeling, especially for small-size cooks and for small medium enterprises. Therefore, the produced design is small-size to provide a reliable portable machine. The right material is also important to ensure that it is not too heavy and easy to use for any category.

An interview and an observation on how to manually peel and separate the peanut skin were performed before the design process is proceeded. The objective tree method was then generated with certain subobjectives. The engineering characteristics of the design were then analyzed by using quality function deployment. The next step was to decide whether designs 1, 2, or 3 fit with the user's needs. The result from weighted objective analysis shows that design 2 is the best design because of the higher score obtained. Design 2 demonstrates that the importance of small-scale machine with fewer components used is appropriate for the skin peeling task.

References

1. Charts PC (1900) The peanut industry. Sci Am 82:243–243. https://doi.org/10.1038/scientificamerican04211900-243a
2. Ugwuoke IC, Okegbile OJ, Ikechukwu IB, John RT (2014) Design and development of manually operated roasted groundnut seeds peeling machine 2:30–33
3. Ogunwole OA (2013) Design, fabrication and testing of A (Manually and electrically operated) roasted groundnut decorticating machine 14
4. Ravindra A, Rohit G, Saurav A (2018) Design and fabrication of groundnut shelling and separating. pp 184–186
5. Soyoye B, Akingba O (2015) Development and performance evaluation of a roasted groundnut Dehulling machine 2:667–675
6. Jambhulkar SG, Handa CC, Kapgate EKM, Tech M (2008) A review on design and development of coconut deshelling machine. Int Res J Eng Technol 5:2137
7. Gana IM, Mohammed KM, Shehu AA, Tauheed IM (2018) Development of a roasted groundnut skin peeling machine. Food Nutr J 8. https://doi.org/10.29011/2575-7091.100093

Chapter 27
Development of a Monitoring System for a Glove Packing Machine During Real-Time Operation Using an Infrared Sensor

M. Asmadi, Malik M. Ali, M. Rezal, and M. N. Norramlee

Abstract An automatic glove packaging system is developed by using a conveyor and pneumatic system to overcome the classical manual approach. The sensors are used to count the quantity of the glove and detect the box. Then, it will send the signals to the control unit. The control unit used is OMRON PLC-CP2E-E30DR-A, which sends an appropriate signal to the pneumatic cylinder to push the glove into the box. In this design, an infrared sensor 1 is used to count the quantity of the glove at the conveyor and the conveyor will stop after it finishes counting. The infrared sensor 2 will detect the box and stop the conveyor. The pneumatic system is built to push and hold the glove at the trap door to the box for packaging. The 3-phase AC motor is used at the belt conveyor. An experiment is done to choose the best delta connection or star connection to supply the current and the power to the conveyor motor. The downscale prototype of the machine is developed to present the function of the machine before building the real industrial machine. Using this system completely eliminates manpower, reduces inspection time, and also increases the production rate.

Keywords Glove packing machine · Infrared sensor · Downscale prototype

---

M. Asmadi · M. M. Ali · M. Rezal (✉) · M. N. Norramlee
RF Laboratory, Universiti Kuala Lumpur, Malaysian Spanish Institute Kulim Hi-Tech Park, 09000 Kulim, Kedah, Malaysia
e-mail: mrezal@unikl.edu.my

M. Asmadi
e-mail: asmadi@s.unikl.edu.my

M. M. Ali
e-mail: amalik@unikl.edu.my

M. N. Norramlee
e-mail: norramlee@unikl.edu.my

T. A. A. Razak et al. (eds.), *Progress in Engineering Technology VI*, Advanced Structured Materials 229, https://doi.org/10.1007/978-3-031-81517-1_27

27.1 Introduction

A glove packaging machine is designed that employs a conveyor and pneumatic system to automatically pack gloves as opposed to utilizing a manual system. This research aims to determine whether the procedure of packaging and collecting gloves is monitored in real time. This project oversees performing continuous monitoring of the glove packaging and collection processes in real time. It is vital to examine the state of the glove prior to packaging it away. To demonstrate the packaging glove machine, a downscale prototype is developed. There are several components to the design, including a cylinder for pushing the glove into the box. The design is comprised of a few different elements, such as a cylinder that is used to force the glove into the box and conveyor belting that is used to move the rubber glove from the operator's side. The suitable component for the system is very important to cover the sequence flow glove packaging. The infrared sensor is applied for sensing the number of gloves and detecting the box. The software CX-programmer is used to create the system to follow the sequence for packing the glove into the box. The only thing that has not been looked at yet is how well the box is sealed after the cylinder has put the rubber glove in it, as well as the quality of the glove itself. Before moving on to the next stage, which is to place the rubber glove box inside the carton box, the operator will personally inspect the quality of the glove, and there is a seal box portion that will guarantee that the box is properly sealed.

27.2 Conveyor System

This project requires machine and technology research. This will assist us to understand challenges and find solutions for our project. Belt conveyors transfer goods, people, and materials. Belt conveyors use belts instead of chains, spirals, hydraulics, etc. This method stretches a flexible material loop between electrically driven rollers. The belt material depends on the system and what is being transported. Usually, a rubber or polymer belt. A belt conveyor is a device used to move physical objects including goods, people, and materials from one location to another. Belt conveyors transport the goods by means of a belt, as opposed to alternative conveying methods that use chains, spirals, hydraulics, etc. A flexible material loop is stretched between electrically driven rollers in this process. The belt conveyor example is used in industry. Since the nature of the things being moved varies, so does the belt material according to the system it is used in. It frequently arrives as a rubber or polymer belt. [3]. Belt conveyors transfer goods, people, and materials. Belt conveyors use belts instead of chains, spirals, hydraulics, etc. To perform efficiently, belt conveyor system components must be carefully constructed. Each conveyor belt has a specified size and type to fit the items, which are loaded in a specific order. Head pulleys have the largest diameter. Drive pulleys can sometimes be many. The discharge end's head pulley is the drive roller with the biggest diameter. A simple belt conveyor's tail pulley is mounted on grooved guides to accommodate belt tension. Later example, various belt conveying systems use a take-up roller to tighten the belt. However, the belt is arguably the most important component. Tension and strength are crucial since the belt is abused during loading and transport. Despite the costs, the demand for longer carrying lengths has fuelled research into new materials. Based on weight, height, and distance, the frame will vary. Simple cantilever-like construction is possible. Trusses can be used for heavier loads. Simple, lightweight chores require aluminium extrusions. To work swiftly and safely, choose the drive motor and pulleys. The tension rolls guide and support the belt. For a traction motor, the torque is used to move a vehicle from a standstill and when the vehicle has started moving it already has some inertia, thus a lower torque is required to propel the vehicle. The series motor has the highest starting torque compared to other types of motors. This is good for other types of the applications except for motor traction. The high starting torque will cause jerks during start-up which are unwanted. To overcome the jerk effect while start-up, the torque of a series motor has to be reduced during start-up. This method stretches a flexible material loop between electrically driven rollers. The belt material depends on the system and what is being transported. Usually, a rubber or polymer belt. A belt conveyor has a conveyor belt, driving motor, pulleys, idlers, tension rolls, and other parts depending on the system. The conveyor belt moves items while the drive motor powers it. Pulleys, idlers, and tension rolls support, tension, and track the belt.

27.3 Methodology

In this research, the tension, speed, and carry load are related to the calculation below, the research utilizing various publications [1–9]. Assembly, part, calculation, idea, and system design of the glove packaging machine's primary components are explained in this section. SolidWorks was used to create 2D and 3D component models for the drawings. The prototype will enhance safety and mechanical system design. The design calculations determine the drawings' dimensions. PLC Omron CP2E controls the glove packaging machine with 18 inputs and 12 outputs. For a simple horizontal conveyor as shown in Fig. 27.1, the effective pulling force is given by Eq. (27.1).

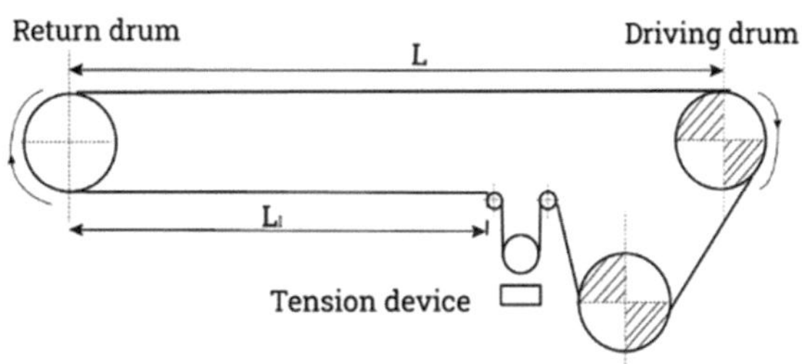

Fig. 27.1 Horizontal conveyor

$$\boldsymbol{F}_{\mathbf{u}} = \boldsymbol{\mu}_{\mathbf{R}} \times \mathbf{g} \times (\mathbf{m} + \mathbf{m}_{\mathbf{b}} + \mathbf{m}_{\mathbf{R}}) \tag{27.1}$$

where F_u is the effective pulling force, μ_R is the friction coefficient when running over the roller, g is the acceleration due to gravity, m is the mass of goods conveyed on the whole length of the conveyor, m_b is the mass of belt, and m_R is the mass of all rotating rollers minus the mass of the drive roller. The speed of the conveyor will be the circumference of the drive pulley multiplied by the revolutions per unit time. Belt conveyors move commodities, products, and people. Figure 27.2 shows the related design of conveyor mapping to the calculation state in this paper.

$$\mathbf{V_C} = \mathbf{D.F} \tag{27.2}$$

where Vc is the speed of conveyor belt, D is the diameter of drive pulley in metres, and F is the revolutions of drive pulley per second. The glove packing design allows it to be run to the box. The conveyor system features a rod to hold the box while the conveyor is running. A functional light tower is needed to monitor the machine while it is running. Figure 27.3 shows the 3D conveyor that transports gloves and boxes.

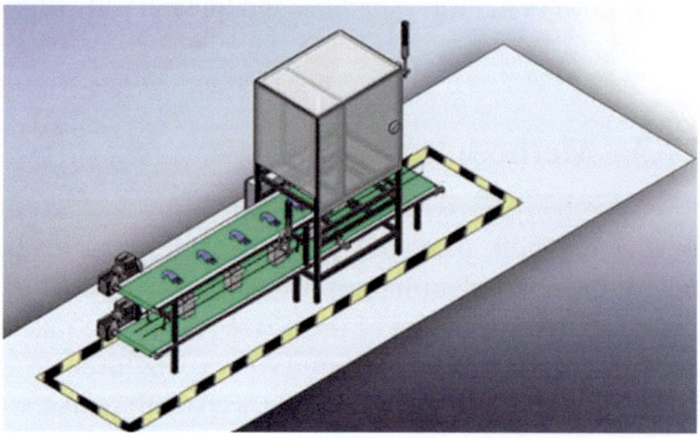

Fig. 27.2 Design structure of glove packing machine

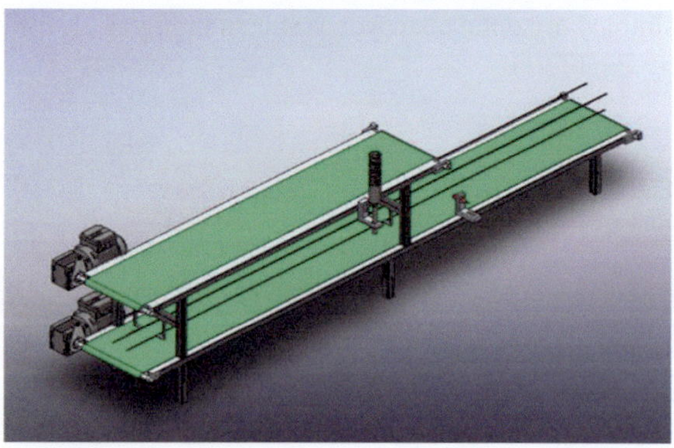

Fig. 27.3 3D conveyor complete design

The complete fabrication design is shown in Fig. 27.4. It shows the assembly view of the prototype after all parts were printed. This prototype is scaled down by a factor of 3 from the real machine to be presented to the company. Figure 27.5 shows the final design of the prototype for the glove packaging machine. The position of the component is located at the machine. The additional part is for the safety of the operator at the machine.

Fig. 27.4 Full view of prototype

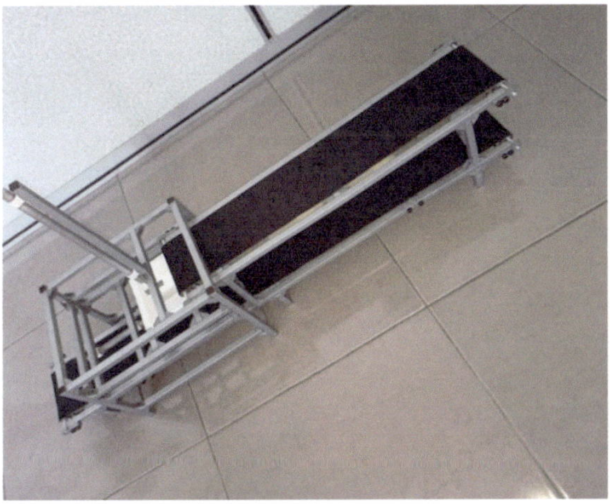

Fig. 27.5 Final design of prototype

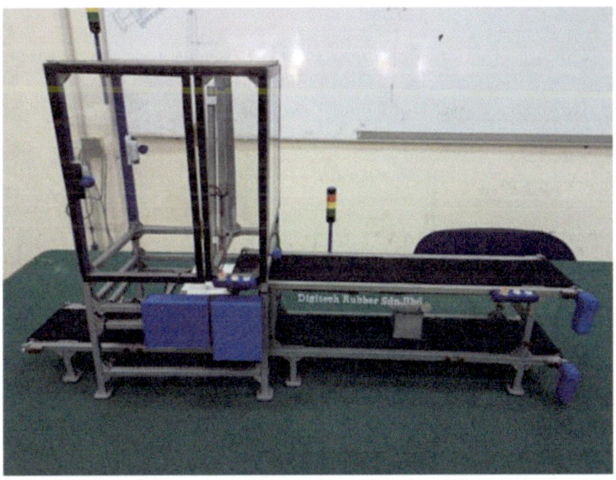

27.4 Result and Discussion

Figure 27.6 shows the delta and star power differences of 0.005 kW to 0.010 kW at 60 A while Fig. 27.7 for 70 A. Figure 27.8 shows the power consumption but the star connection is steady after beginning at 10–15 s, unlike the delta connection, which is stable at the start but unstable at 90 A.

The delta connection for motor conveyors is more efficient, reliable, and cost-effective than star connections, according to an experiment and analysis. Delta

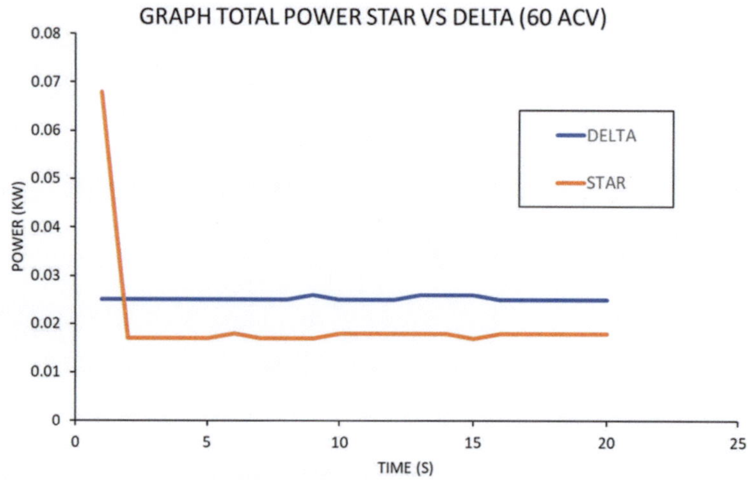

Fig. 27.6 Power consumption comparison 60 A

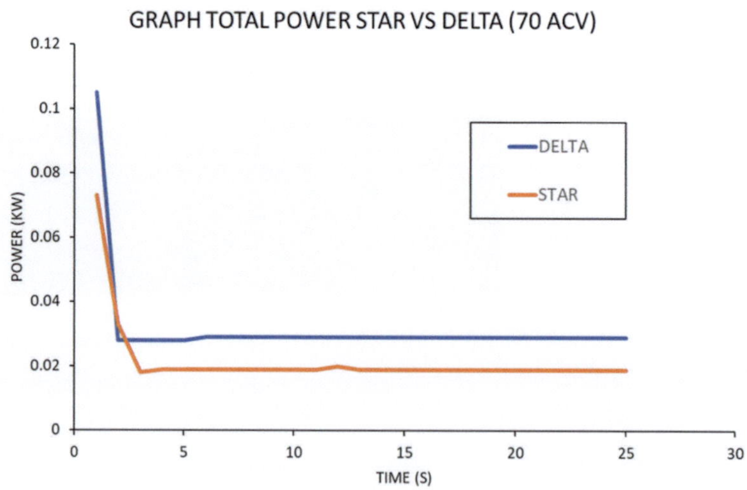

Fig. 27.7 Power consumption comparison 70 A

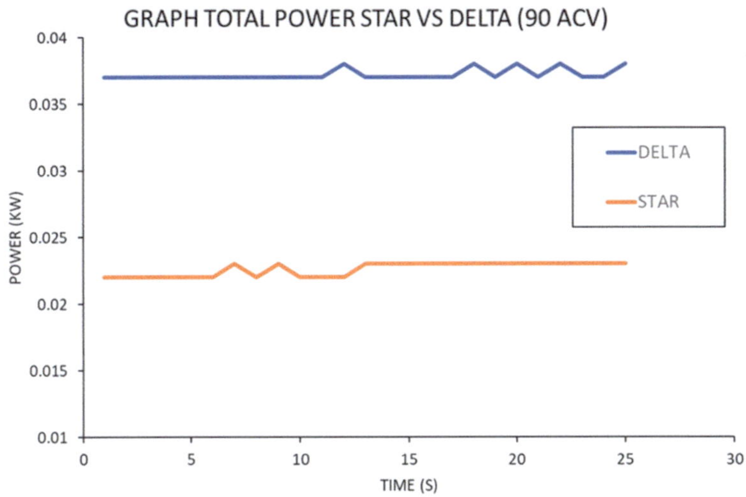

Fig. 27.8 Power consumption comparison 90 A

connections reduce motor and control system noise and allow better temperature management. Balanced delta connections require three-phase power, compared to star connections which use one power route. This downscale prototype is designed as a reference for the company Digitech Rubber Sdn. Bhd. This design is suitable for packaging needs where only small places are available. The pneumatic system is using the lead screw method to present the real function to push the glove and the trap door as shown in Fig. 27.9.

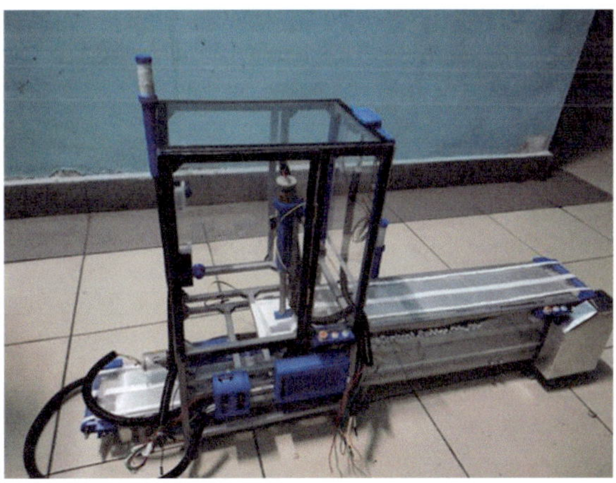

Fig. 27.9 Downscale prototype with real movement

27.5 Conclusions

In conclusion, this machine provides a low-cost, efficient box-filling machine. Before establishing the machine for glove factory industrial production, the prototype might be improved to achieve the desired performance. This machine is an inexpensive, efficient box-filling machine. The system's internal components are used and a controller is constructed to simulate the packing glove machine's prototype. This ensures equipment functionality. Thus, a packing glove machine model is possible. Before building the glove factory's industrial machine, the prototype might be modified to get the needed performance. These changes may improve the performance.

References

1. Bortnowski P, Kawalec W, Król R, Ozdoba M (2022) Identification of conveyor belt tension with the use of its transverse vibration frequencies. Meas J Int Meas Conf 190. https://doi.org/10.1016/j.measurement.2022.110706
2. Noroozi M, Radiman S, Zakaria A (2014) Influence of sonication on the stability and thermal properties of Al_2O_3 nanofluids. J Nanomater 2014:1–10. https://doi.org/10.1155/2014/612417
3. Draganová K, Semrád K, Spodniak M, Cúttová M (2020) Innovative analysis of the physical-mechanical properties of airport conveyor belts. Trans Res Procedia 51:20–27. https://doi.org/10.1016/j.trpro.2020.11.004
4. He D, Pang Y, Lodewijks G (2016) Speed control of belt conveyors during transient operation. Powder Technol 301:622–631. https://doi.org/10.1016/j.powtec.2016.07.004
5. Li Z, Zhu X, Zhou J (2020) Intelligent monitoring system of coal conveyor belt based on computer vision technology. In: Proceedings–2019 6th international conference on dependable systems and their 24 applications, DSA 2019, pp 359–364. https://doi.org/10.1109/DSA.2019.00055
6. Liu W, Wan P, Cheng J, Ma Y, Jing C (2022) High-precision position control of belt drive system based on OPC communication. Int J Adv Manuf Technol 122(1):1–10. https://doi.org/10.1007/S00170-021-07859-W
7. Pang Y, Lodewijks G (2011) Improving energy efficiency in material transport systems by fuzzy speed control. LINDI 2011–3rd IEEE international symposium on logistics and industrial informatics, Proceedings, pp 159–164. https://doi.org/10.1109/LINDI.2011.6031139
8. Fedorko G, Molnar V, Marasova D, Grincova A, Dovica M, Zivcak J, Toth T, Husakova N (2013) Failure analysis of belt conveyor damage caused by the falling material. Part II: application of computer metrotomography. Eng Fail Anal 34:431–442. https://doi.org/10.1016/J.ENGFAILANAL.2013.09.016
9. Fedorko G, Molnar V, Marasova D, Grincova A, Dovica M, Zivcak J, Toth T, Husakova N (2014) Failure analysis of belt conveyor damage caused by the falling material. Part I: experimental measurements and regression models. Eng Fail Anal 36:30–38. https://doi.org/10.1016/j.engfailanal.2013.09.017.062

Chapter 28
Development of a Smart Monitoring Hydro Turbine Installation in a Water Pipeline

Noor Iman Fitri bin Abu Bakar,
Muhammad Afif Asnawi bin Muhammad Azizan,
Muhammad Rahimy bin Mohd Radzi, Wan Nur Suhada Wan Mohd Zor,
and Ahamad Zaki Mohamed Noor

Abstract Researchers have implemented measures to improve the energy structure in response to the climate agreement by decreasing the proportion of coal-fired thermal power and increasing the use of clean energy. But because there are so few large-scale energy storage facilities, wind and solar electricity continue to have low utilization rates. The utilization of water from an underground pipeline is suggested in this research. After installing a turbine, the water flow from the pipeline can provide a significant amount of electricity. This study also describes the creation of a hydro-turbine system's intelligent Internet of Things monitoring system. The micro-hydro water inflow is still controlled manually as of right now. This fact is problematic since it requires too many human resources and, when carried out manually, the accuracy level is incorrect. The prior monitoring method could only be observed near the generator, remote monitoring was not possible. It is made to allow automatic water entry regulation via a smartphone application to get around the issues. The monitoring system is also created and programmed so that it can be viewed using a smartphone application at any time and from any location. The technique is an experimental

N. I. F. A. Bakar · M. A. A. M. Azizan · M. R. M. Radzi · W. N. S. W. M. Zor
Mechanical Engineering Section, Universiti Kuala Lumpur, Malaysian Spanish Institute Kulim Hi-Tech Park, 09000 Kulim, Kedah, Malaysia
e-mail: iman.abu23@s.unikl.edu.my

M. A. A. M. Azizan
e-mail: afif.azizan25@s.unikl.edu.my

M. R. M. Radzi
e-mail: rahimy.radzi@s.unikl.edu.my

W. N. S. W. M. Zor
e-mail: suhada.zor@s.unikl.edu.my

A. Z. M. Noor (✉)
Manufacturing Section, Universiti Kuala Lumpur, Malaysian Spanish Institute Kulim Hi-Tech Park, 09000 Kulim, Kedah, Malaysia
e-mail: ahamadzaki@unikl.edu.my

T. A. A. Razak et al. (eds.), *Progress in Engineering Technology VI*, Advanced Structured Materials 229, https://doi.org/10.1007/978-3-031-81517-1_28

technique that produces a water turbine design as well as prototype-scale experiments. The type of turbine employed consisted of 20 types with blade diameters of 50–40 mm and pipe diameters of 55 mm. Three l/min was the inflow velocity. To examine the impact of various numbers of blades on the turbine movement and flow pressure, the experiment is carried out using 20 different numbers of blades, ranging from 2 to 6. Using a sophisticated monitoring system, the performance of each blade was tracked. The findings demonstrate that the system that has been developed can perform admirably. The outcomes demonstrated that the optimum blades for moving because of the water flow from the intake area were those with five blades. Meanwhile, the turbine's RPM decreased when it had blades 2–4 and 6. Because the angle of the guiding vane is too tiny, there will be turbulence, making the blade design unsuitable. The outcomes also demonstrated that the 5 blades outperformed in terms of movement velocity and flow pressure.

Keywords Hydroturbine · Blades · Renewable energy · Turbine · Generator · IOT

28.1 Introduction

Nowadays, hydropower turbine systems may be used to recover energy from networks by installing a brake pressure tank in water supply networks to address high flow pressure. Additionally, researchers demonstrated the first technical and financial viability of the strategy to reduce energy use [1]. Because it influences a turbine's speed and efficiency, the number of blades is crucial. The swept area of a wind turbine's blades directly relates to the power it generates from the wind; as a result, the blades directly impact power generation [2]. The longer the blade, the more energy the turbine will create, but it must be balanced by at least two other objects. To integrate the hydroponic system, a communication protocol namely the Internet of things (IoT) needs to be embedded since it is suitable for many applications [3]. The management of the production process has altered because of technological advancements. The ability to outfit the clean energy industry with Internet of Things (IoT) technology has made it possible for equipment to link and communicate with one another to accelerate the process [4]. The booming of the high advance technology give impact to the manufacturing industry during German strategy initiative to lead market and provider of advanced manufacturing solution that called Industry 4.0 [5].

28.2 Literature

Hydro turbines are used in hydroelectric power facilities to convert the kinetic energy of flowing water into rotational energy for a generator shaft. When water is fed to the turbine's blades, the turbine rotates or spins. To generate electricity from water, a technique known as hydropower, these turbines are crucial [6]. For almost thirty years, researchers have been studying the cost optimization of piped water networks. Numerous restricted optimization methods, ranging from Genetic Algorithms to Linear Programming, have been utilized to address different iterations of the cost optimization issue.

28.2.1 Effect of Blade Number on Torque

The relationship between the torques and different turbine blades counts (2–6 blades) will show a different result by no doubt that the turbine with the five-blade configuration generated the most torque. According to Sriram et al. [7] five-bladed turbines are the most effective at absorbing the effects of water flow. It is also clear that when the number of blades increased—to five—the surface area touching the water increased as well, resulting in a larger torque. Additionally, resistance prevented the next blade in the series from moving, which decreased the torque the turbine produced because of water flowing through the hole.

28.2.2 Electrical Smart Monitoring System

Research conducted by Akib et al. [6] claims that because of the world's ongoing technological development, there will be a greater need for energy in the future. The author researched how hydropower will be generated from those sources and how it can power a home. The authors also mentioned that users will be able to track the energy generated and use it as needed. Other researchers Pasika et al. [8] the IoT was examined for smart, cost-effective water quality monitoring systems. A system comprised of multiple sensors was provided by researchers to assess a variety of results and characteristics.

28.2.3 Angle of Blade

Masud et al. [9] by adjusting the angle of the blade inclination, it was possible to see differences in how the angle affected the hydrokinetic turbine's efficiency. The findings indicate that as compared to the 90-degree inclination turbine, the 30-degree inclination turbine has a very high-performance tendency with an average rotational velocity. The efficiencies change with the angle of inclination. With a high efficiency of 58% at 0.78 (Nm) torque, a turbine with a 30-degree inclination achieves an outstanding performance. The highest efficiency of a 90° inclination turbine is 27% at 0.48 (Nm) torque.

28.3 Methodology

Figure 28.1 shows the flowchart for this project. All the components for the prototype were assembled. Install the turbine in the middle of a straight pipe. A water pump installed at the pipe's intake would force water through it. To store the electricity created by the blade, it will be connected to the motor and transmitted to the box located outside of the aquarium. Two water flow sensors will be installed in the piping between the water pump and the turbine to track the flow of water. The second flow meter is going to be installed on the pipe's endpoint. NodeMCU ESP8266, due to its compact size and user-friendliness, has been selected as the microcontroller for this project's coding system.

28.3.1 Experimental Set-Up/Design Approaches

Table 28.1 shows blade design criteria. For this project's prototype, the pipe diameter is 55 mm. This means that the diameter of each blade must be between 40 and 50 mm (between 55 mm), the thickness of the base must be at least 10 mm to withstand the force of the water, and the blades themselves must have an optimal shape to maximize the flow of water from the pipe to the tip. Class D PVC pipe is ideal since it can be quickly assembled and set up for experiments.

The blade design that will be used in this project is shown in Table 28.1. There will be 20 samples altogether for prototype purposes. Four major criteria will be chosen which are the angle, size, hub size, and number of blades. This will be proportional to the size of pipe that is used for the prototype. Two inches PVC pipe size is chosen as it is the most suitable size to tally with the waterflow sensor DN50-YSF201 which is also 2-inch water flow sensor for the Arduino.

28.4 Result and Discussion

The 20-blade experiment arrangement used in this chapter will be used to draw conclusions. The speed of the flow and the blade's rotation is considered. The monitoring system at Blynk and Thinkspeak was used to collect all the data.

Three aspects need to be considered to select the best blade which are outlet flow rate, turbulence flow, and velocity magnitude during outlet flow. The best blade should have high outlet water flowrate, less turbulence flow, and uniform velocity magnitude at water outlet in this research.

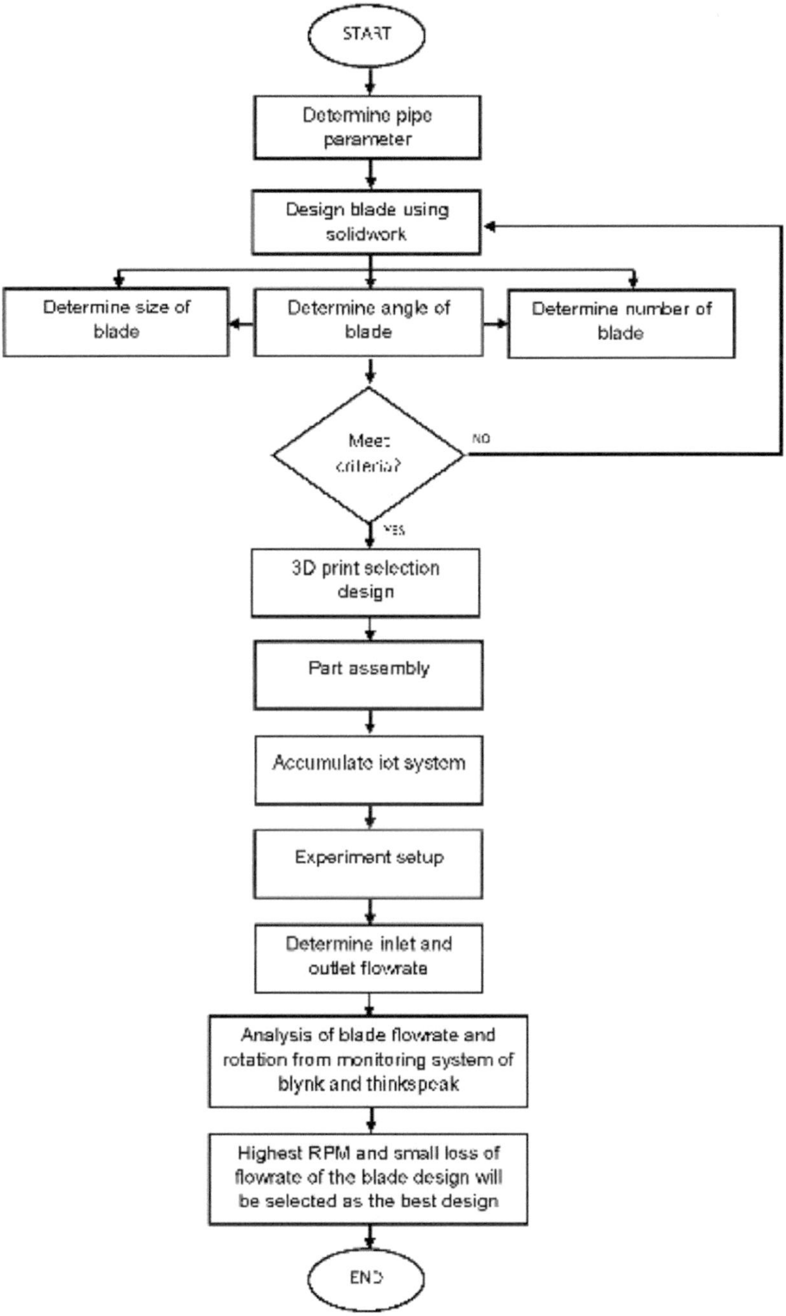

Fig. 28.1 Flowchart of hydroturbine performance experiment

Table 28.1 Blade design criteria

Angle	Size (mm)	Hub size (mm)	Blade thickness (mm)	No. of blades
90	50	20	5	2
				3
				4
				5
				6
45	40	15	2.5	2
				3
				4
				5
				6
			Total turbine	20pcs

28.4.1 Blades Inlet and Outlet Flowrate

Table 28.2 shows flowrate result of 20 blades. All the blades were successfully run and pass the test. At inlet flowrate, with initial flowrate of 3.0 L/minutes, there were only two blades that interrupt the water flowrate in the pipeline. It is big 90 2b which has the slowest flowrate of 2.802 L/min and small 45 5 b which has the flowrate of 2.924 L/minutes. There is a factor like the reverse flow when the turbine rotates in the pipeline that causes the flow sensor in the hall turbine to be interrupted by the reverse flow that may have caused the flowrate to be a little bit slower than the initial flowrate.

28.4.2 Rotation

Table 28.3 shows the rotational result of 20 blades. There are only 8 blades that have successfully made a rotation from the water flow velocity and 12 blades that cannot make a single rotation from the water current. The blade with 90° angle for all criteria cannot produce a single rotation. A major factor is that the blade cannot capture enough water to make it rotate and the water also cannot push the blade and give enough pressure to force the blade to rotate the highest rotational per minute (RPM) that the blade design produces was small 45 5b which has 5 number of blades.

The optimum number of hydroturbine should have been 5–6 number of blades which give the optimum torque to generate electricity. The higher the number of blades, the higher the torque it produces but in exchange the flow of water will be much.

Table 28.2 Flowrate result of 20 blades

No	Blade	Inlet (l/m)	Outlet (l/m)
1	Big 90 2b	3.046	3.655
2	Big 90 3b	3.046	3.046
3	Big 90 4b	3.168	3.289
4	Big 90 5b	3.046	2.924
5	Big 90 6b	2.802	2.802
6	Big 45 2b	3.168	3.046
7	Big 45 3b	3.168	3.168
8	Big 45 4b	3.168	3.106
9	Big 45 5b	3.046	3.046
10	Big 45 6b	3.046	3.168
11	Small 90 2b	3.046	3.168
12	Small 90 3b	3.168	3.168
13	Small 90 4b	3.168	3.046
14	Small 90 5b	3.046	3.289
15	Small 90 6b	3.168	3.046
16	Small 45 2b	3.046	3.168
17	Small 45 3b	3.168	3.046
18	Small 45 4b	3.046	3.168
19	Small 45 5b	2.924	3.046
20	Small 45 6b	3.168	3.046

28.5 Conclusions

This research concluded that a turbine with five blades is optimal because it generates the maximum rotations per minute when subjected to the impact of water flow. Both the number of blades and their arrangement are crucial to the turbine's performance. However, the slipstream effect worsens with increasing blade counts. Reduced performance and efficiency are due to insufficient blade count. Having too many blades adds extra bulk and money to the manufacture. Having the right number of blades for the generator performance curve is crucial for maximizing the turbine performance and efficiency.

In this study, an IoT-based monitoring system for a 50 mm hydroturbine is presented. Data from the Internet of Things monitoring system are primarily used to optimize hytdroturbine operations. The data from the IoT monitoring system's voltage, current, and various sensors are sent to an Arduino Uno board for processing. Next, the information is transmitted to a development board, where it may be viewed on a screen, uploaded to a website, and saved for later use.

Table 28.3 Rotation result of 20 blades

No.	Blade	Rotation (RPM)
1	Big 90 2b	0
2	Big 90 3b	0
3	Big 90 4b	0
4	Big 90 5b	0
5	Big 90 6b	0
6	Big 45 2b	0
7	Big 45 3b	173.475
8	Big 45 4b	167.775
9	Big 45 5b	171.3
10	Big 45 6b	191.25
11	Small 90 2b	0
12	Small 90 3b	0
13	Small 90 4b	0
14	Small 90 5b	0
15	Small 90 6b	0
16	Small 45 2b	0
17	Small 45 3b	181.95
18	Small 45 4b	182.4
19	Small 45 5b	202.65
20	Small 45 6b	175.8

References

1. Chen H, Kan K, Wang H, Binama M, Zheng Y, Xu H (2021) Development and numerical performance analysis of a micro turbine in a tap-water pipeline. Sustainability (Switzerland)
2. Adeyeye KA, Ijumba N, Colton J (2021) The effect of the number of blades on the efficiency of a wind turbine. IOP C SER EARTH ENV
3. Hafidz MMF, Damanhuri AAM, Hambali RH et al (2020) Recent applications of internet of things (IoT) in manufacturing sector—a review. Int J Sci Technol Res 9(9):334–339
4. Fauadi MHFM, Damanhuri AAM, Hambali RH, Mohamed AZ, Noor NIA (2020) Recent applications of internet of things (IoT) in manufacturing sector–a review. Int J Sci Technol Res 9(09):334–339
5. Noor AZM, Dzulkafle MS, Bohri NM, Khaireez MHM, Jupri SS, Fauadi MHFM (2023) Development of smart agriculture (Smart Hydroponic) to monitor soil humidity level. In: Abu Bakar MH, Abdul Razak TA, Öchsner A (eds) IT applications for sustainable living. SpringerBriefs in Applied Sciences and Technology. Springer, Cham, pp 25–34
6. Akib et al (2021) International conference on innovation and intelligence for informatics, computing, and technologies (3ICT)
7. Sritram P, Suntivarakorn R (2019) The effects of blade number and turbine baffle plates on the efficiency of free-vortex water turbines. IOP C SER EARTH ENV
8. Pasika S, Gandla ST (2020) Smart water quality monitoring system with cost-effective IoT. Heliyon 6(7):e04096
9. Masud IA, Suwa Y (2018) Effect of blade inclination angle on the efficiency of hydrokinetic turbine in an undershoot zero head system. Int J Mat Mech Manufactur 6(6):388–391

Chapter 29
Theoretical Calculation of the Tree Structure Height Limit by WKB Approximation of the Schrödinger Equation

John Jabaraj Devadason

Abstract We analyse the limit to tree height by employing elementary quantum mechanics. The xylem is assumed as a one-dimensional vertical potential well without any rigid walls; transporting a water molecule in an oscillating medium. A macromolecule with coherence can behave quantum mechanically. By applying the quantization rule of the Wentzel–Kramers–Brillouin approximation of the Schrödinger equation to the Bernoulli equation, a general quadratic equation on the limit of the tree height is derived. The limit of tree height is therefore calculated to range from 0 to 203.87 m, and approaches the peak value of 101.94 m (as the quantum number increases).

Keywords Tree height structure · Quantum mechanics · Potential wells · Bernoulli equation

29.1 Introduction

The sheer size and height of some millennium-old trees are astounding. The tallest living tree known today is a Coast Redwood (sequoia sempervirens), and it stands at ~ 116 m [1]. This tree, discovered and named as the Hyperion in 2006, is only about 600 years old (considered quite young by Coast Redwood standard) and is still growing robustly. Nevertheless, other taller trees have been historically described and measured. The tallest tree ever mentioned in literature is mountain ash (eucalyptus regnans) at an unverified 160 m [2]. The usual question that invariably arises when studying tall trees is how much more can they grow. Is there a certain tree height limit?

J. J. Devadason (✉)
Universiti Kuala Lumpur, Malaysian Spanish Institute, Kulim Hi-Tech Park, 09000 Kulim, Kedah, Malaysia
e-mail: john@unikl.edu.my

T. A. A. Razak et al. (eds.), *Progress in Engineering Technology VI*, Advanced Structured Materials 229, https://doi.org/10.1007/978-3-031-81517-1_29

29.2 Literature Review

29.2.1 Past Studies on Tree Height Limit

The limit to tree height has been studied widely in the past through various methods. By modelling the structure and allometry of trees, the maximum tree height is determined to be 120 m [3]. Meanwhile, a regression analysis of the height gradients in leaf functional characteristics, estimated the maximum height of Coast Redwood trees to be 122–130 m [4]. In another study, the water transportation across pits between tracheid in the Douglas-fir (pseudotsuga menziesii) trees was determined as almost zero at heights of > 109 m [5].

Some probable factors responsible for the existence of tree height limits have been proposed based on some constraints. The hydraulic constraint reflects the battle between gravity and transpiration in the transportation of water and dissolved mineral salts to the leaf for photosynthesis [6]. Meanwhile, the mechanical constraint considers the structural instability that increases with the height of trees [7]. Nonetheless, there is also the genetic limitation that preexists in all tree species which intrinsically determines its general height [8]. Even then, the height of trees of a certain species can vary due to the constraint on resources [9].

The maximum height of trees changes continually as the tallest living tree known today is still growing robustly [1]. There may as well be other taller trees waiting to be found in future. Past studies have estimated the limit to tree height and they utilized classical methods. Here in our study, we offer an alternative and interesting quantum mechanical view on the limit to tree height. Nevertheless; in order to apply quantum mechanics to a macroscopic system, coherence or entanglement is required [10]. Coherence is present in the xylem [11].

29.3 Methodology/Experimental Set-Up

First, we introduce some assumptions concerning the xylem, the water molecule, and the applicability of quantum mechanics. The xylem is then modelled here as a vertical potential well without rigid walls in which a fluid element moves according to the Bernoulli equation. Consequently, the equation on the tree height limit is then derived by utilizing the quantization rule of the Wentzel–Kramers–Brillouin (WKB) approximation of the Schrödinger equation and then further analysed.

29.3.1 Modelling the Xylem

Several assumptions were made in modelling the xylem and the elements that travel within.

(a) The Bernoulli equation is applicable in the xylem

The xylem transports water and dissolved mineral salts from roots to the leaf for the production of photoassimilates during photosynthesis [12]. Hence, the xylem is assumed as a fluid system where the Bernoulli equation can be applied. The Bernoulli equation for point 1 and point 2 of a fluid system is

$$P_1 + 0.5\rho v_1^2 + \rho g l_1 = P_2 + 0.5\rho v_2^2 + \rho g l_2 \quad (29.1)$$

where P_1 and P_2 are the pressures at the chosen point 1 and point 2. The fluid flow speeds therein are v_1 and v_2, while the elevations above a reference plane are l_1 and l_2. The g is the acceleration due to gravity (9.81 m/s^2) while ρ is the density of the fluid at all points.

(b) The xylem is assumed as a potential well without any rigid walls, traversed by the water molecule.

The xylem is an open channel from root to leaves. During transpiration, the gas microbubbles at the inner walls of the xylem oscillate in coherence [11]. Thus, even though the xylem is a macroscopic system, coherence exists within it. Coherence is required in order to observe quantum mechanical behaviour in macroscopic systems [10]. Hence, the xylem is assumed here as a one-dimensional vertical potential well, $V(L)$ without any rigid walls. Furthermore, the macroscopic water droplet subjected to an oscillating medium can be induced to behave as a quantum mechanical particle [13]. The corresponding quantization rule; with momentum in the y-axis as $p(L)$, h as the Plank's constant and quantum number as $n = 0, 1, 2, 3, \ldots$ is [14]:

$$\int_{l_2}^{l_1} p(L)\mathrm{d}L = (n + 0.5)\pi h \quad (29.2)$$

29.3.2 Application of Quantum Mechanics

The Bernoulli equation is a variant of the energy conservation theorem describing a macroscopic phenomenon of fluid flow, which is derived by a direct application of Newton's second law to a fluid particle moving along a streamline [15]. A macroscopic water droplet subjected to an oscillating medium has been shown to behave as a quantum mechanical particle [13]. Meanwhile, the quantization rule of WKB (a simple form of quantum mechanics) describes the probability of finding a particle in a specific potential energy landscape. Hence, the application of the quantization rule of WKB to the Bernoulli equation is justified.

First, the Bernoulli Eq. (29.1) is modified to obtain Eq. (29.3). The point 1 is assigned at the base of the tree while point 2 is the highest tree crown. The height of the tree, L is the vertical difference between the point 2 and point 1 is $(l_2\_l_1)$. Therefore, $L = l_2$ as l_1 can be assumed 0 m. Furthermore, the velocity of fluid at the base of the tree is assumed ~ 0 m/s; as the root system acts as a reservoir, while at the highest tree crown is set as $v_2 = v$. The pressure difference between point 1 and point 2 is $\Delta P = P_2 - P_1$.

$$
\begin{aligned}
P_1 + 0.5\rho v_1^2 + \rho g l_1 &= P_2 + 0.5\rho v_2^2 + \rho g l_2 \\
0.5\rho(v_1^2 - v_2^2) &= (P_2 - P_1) + \rho g(l_2 - l_1) \\
-0.5(m/V)v^2 &= \Delta P + (m/V)\mathrm{gL} \\
p &= -\frac{2}{v}(\Delta \mathrm{PV} + \mathrm{mgL})
\end{aligned}
\tag{29.3}
$$

The momentum, p is a function of the maximum tree height (L). Meanwhile, density of fluid element, ρ, is the ratio of its mass (m) to volume (V).

Equation (29.3) is then applied to Eq. (29.2).

$$
\int_{l_2}^{l_1} p(L)\mathrm{d}L = (n + 0.5)\pi h
$$

$$
\int_0^L \frac{-2}{v}(\Delta \mathrm{PV} + \mathrm{mgL})\mathrm{d}L = (n + 0.5)\pi h
$$

$$
gL^2 + 2(\Delta P/\rho)L + (n + 0.5)\pi \mathrm{hv}/m = 0 \tag{29.4}
$$

29.4 Result and Discussion

The velocity of water travelling up the trunk of a tree depends on both the type of tree and the transpirational demand placed on the xylem. For woody species, the maximum flow velocity of water in the xylem is approximately 12 mm/s [16]. The transportation of water upwards requires negative pressure. This is mostly attained by transpiration caused by the evaporation of water at the leaves. The maximum difference of negative pressure possible, between leaf surface and base of a tree, must be less than − 1 MPa [17]. This is because if the pressure difference is > 1 MPa, air bubble cavitation and embolism occur and kill the tree crown [18]. On another note, the mass of a molecule of water is 3×10^{-26} kg while its density is 1000 kg/m^3 [19]. By substituting all these values into Eq. (29.4), a general quadratic equation is obtained as:

$$
gL^2 + 2(\Delta P/\rho)L + (n + 0.5)\pi \mathrm{hv}/m = 0
$$

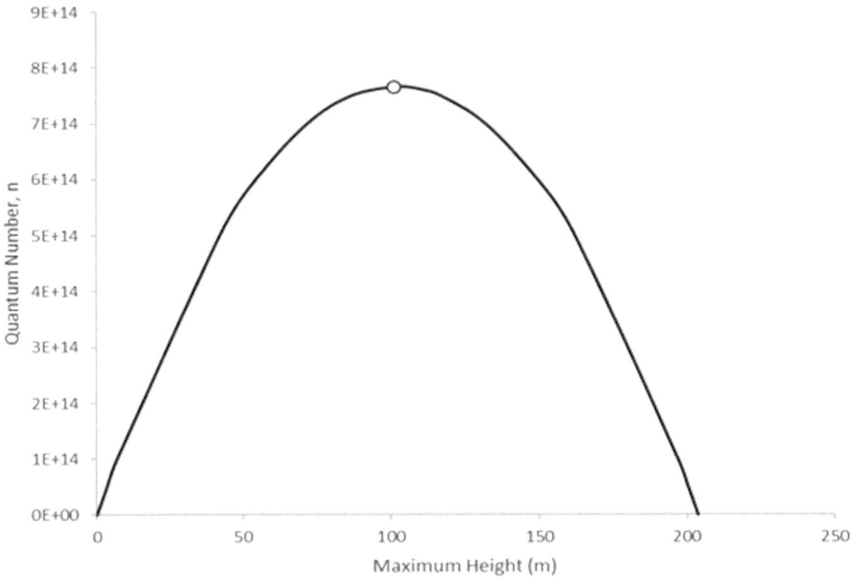

Fig. 29.1 The height limit of tree ranges from 0 to 203.87 m. As the quantum number increases the height limit of tree approaches the value of 101.94 m

$$[9.81]L^2 + 2\left(-1 \times 10^6/1000\right)L + (n + 0.5)\pi h\left[12 \times 10^{-3}\right]/\left[3 \times 10^{-26}\right] = 0$$
$$9.81L^2 - 2000L + 1.33 \times 10^{-10}(n + 0.5) = 0 \tag{29.5}$$

The relationship between the maximum height of tree, L and the quantum number, n is obtained by solving Eq. (29.5), as depicted in Fig. 29.1.

We calculated the limit to tree height by utilizing the elementary principles of quantum mechanics. Specifically, the application of the quantization rule of WKB to the Bernoulli equation (of fluid transportation through the xylem) shows that the tree height limit can range from 0 to 203.87 m. Nevertheless, the quantization rule of WKB approximation becomes more defined with the increase in quantum number [20]. Hence, the tree height limit is found to approach a value of 101.94 m as the quantum number increases (Fig. 29.1).

All past studies on tree height limits obtained some numerical values based on calculations done on some specific constraints or limitations. For example, the tree height limit was calculated to be 109 m [5], 120 m [3] and 122–130 m [4]. In contrast, our study demonstrates that the limit to tree height has a wide range of 0–203.87 m. This calculated range mirrors the nature more closely than the past studies; as the maximum height of plants varies from 0 m upwards, depending on the species. Furthermore, a few tall trees have been historically noted to be 130–160 m [2] which falls within our calculated range. Nevertheless, the upper limit of 203.87 m obtained by our study is yet to be seen in nature. This height may not be achievable due to additional resource constraints. On another note, the average height of the mature trees of Coast Redwood is determined as ~ 100 m [21, 22]. This agrees with the finding of our study where the tree height limit peaks at 101.94 m as the quantum number increases.

Several limitations exist in this study due to the assumptions made. Molecules of water, dissolved mineral salts, and photoassimilates travel within the three-dimensional xylem. However, we treated the xylem as a one-dimensional potential well without rigid walls, containing a molecule of water (subjected to an oscillating medium). The other heavier molecules and the interactions between molecules were not taken into account. The accuracy of this study may be affected by these factors.

29.5 Conclusion

In this paper, we calculated the limit to tree height by application of the quantization rule of WKB to the Bernoulli equation of fluid transportation through the xylem. The result shows that the tree height limit can range from 0 to 203.87 m. Furthermore, the tree height limit is found to approach the value of 101.94 m as the quantum number increases.

References

1. Martin G (2006) World's tallest tree, a redwood, confirmed. SFGate, 2009–09–29
2. Carder A (1995) Forest giants of the world: past and present. Fitzhenry and Whiteside, Ontario
3. West GB, Brown JH, Enquist BJ (1999) A general model for the structure and allometry of plant vascular systems. Nature 400:664–667
4. Koch GW, Sillett SC, Jennings GM, Davis SD (2004) The limits to tree height. Nature 428:851–854
5. Domec JC, Lachenbruch B, Meinzer FC, Woodruff DR, Warren JM, McCulloh KA (2008) Maximum height in a conifer is associated with conflicting requirements for xylem design. PNAS 105(33):12069–12074
6. Friend AD (1993) The prediction and physiological significance of tree height. In: Solomon AM, Shugart HH (eds) Vegetation dynamics and global change. Chapman and Hall, New York
7. Kempes CP, West GB, Crowell K, Girvan M (2011) Predicting maximum tree heights and other traits from allometric scaling and resource limitations. PLoS ONE 6(6):e20551
8. Becker P, Meinzer FC, Wullschleger SD (2000) Hydraulic limitation of tree height: a critique. Funct Ecol 14:4–11
9. Niklas K (2007) Maximum plant height and the biophysical factors that limit it. Tree Physiol 27:433–440
10. Arndt M, Juffmann T, Vedral V (2009) Quantum physics meets biology. HFSP J 3(6):386–400
11. Laschimke R, Burger M, Vallen H (2006) Acoustic emission analysis and experiments with physical model systems reveal a peculiar nature of the xylem tension. J Plant Physiol 163(10):996–1007
12. Holbrook NM, Zwieniecki MA (2011) Vascular transport in plants (physiological ecology). Academic Press, USA

13. Couder Y, Fort E (2006) Single-particle diffraction and interference at a macroscopic scale. Phys Rev Lett 97:154101
14. Zettili N (2001) Quantum mechanics: concepts and applications. John Wiley & Sons, Great Britain
15. Munson BR, Young DF, Okiishi TH, Huebsch WW (2009) Fundamentals of fluid mechanics, 6th edn. John Wiley & Sons Inc., USA
16. Jones HG (1983) Plants and microclimate. Cambridge University Press, Cambridge, UK
17. Zimmermann U, Meinzer F, Bentrup FW (1995) How does water ascend in tall trees and other vascular plants? Ann Bot 76:545–551
18. Pockman WT, Sperry J (2000) Vulnerability to cavitation and the distribution of Sonoran desert vegetation. Am J Bot 87:1287–1299
19. Kumar N (2003) Comprehensive physics for Class XI. Laxmi Publications pvt. Ltd., New Delhi
20. Friedrich H (2002) Semiclassical and large quantum number limits of the Schrödinger equation. In: Proceedings of the fourth international conference on dynamical systems and differential equations, Wilmington, USA, pp 288–294
21. McBride J, Jacobs D (1977) Ecology of redwood and the impacts of man's use of the redwood forest as a site for recreational activities-a literature review. In: Muir woods research project, department of forestry and conservation, University of California, Berkeley. Technical Report No. 1, CX 8000-6-0035
22. Snyder JA (1992) The ecology of sequoia sempervirens: an addendum to "on the edge: nature's last stand for coast redwoods". M.A. Thesis, Department of Biological Sciences: San Jose State University

Chapter 30
Applying the TRIZ Concept to Develop an Innovative Hydrofluorocarbons Fire Suppression Nozzle Alternative

Jum'azulhisham bin Abdul Shukor, Muhammad Al'Hapis bin Abdul Razak, Azmi bin Hassan, and Mohd Fauzi bin Abu Hassan

Abstract The design of the distribution nozzle is important for effective fire extinguishing, but not all types are suitable for all types of gaseous fire suppression agents. To replace hydrofluorocarbon fire extinguishing agents, a new nozzle design is needed that meets specific functions. Before designing the new distribution nozzle, a conceptual design is essential. The ultimate goal of developing a new distribution nozzle for a dedicated agent is to ensure the effectiveness of turbulence flow during extinguishing. To assist in the conceptual design process, the theory of inventive problem-solving method was employed in this research, matching identified contradictions to improve and preserve the desired features. Based on this approach, three design principles were determined and proposed for the fabrication of a dedicated fire suppression distribution nozzle alternative. The results obtained from the matrix were used to propose and establish a better distribution nozzle prototype for the use of a dedicated hydrofluorocarbon fire suppression alternative.

Keywords TRIZ · Hydrofluorocarbons · Clean agent · Fire suppression alternative · Distribution nozzle prototype

---

J. bin Abdul Shukor (✉) · M. A. bin Abdul Razak
Manufacturing Section, Universiti Kuala Lumpur Malaysian Spanish Institute, Lebuhraya Perdana, Kulim Hi-Tech Park, 09090 Kulim, Kedah, Malaysia
e-mail: azulhisham@unikl.edu.my

M. A. bin Abdul Razak
e-mail: alhapis@unikl.edu.my

A. bin Hassan · M. F. bin Abu Hassan
Electrical, Electronics and Automation Section, Universiti Kuala Lumpur Malaysian Spanish Institute, Lebuhraya Perdana, Kulim Hi-Tech Park, 09090 Kulim, Kedah, Malaysia
e-mail: azmi.hassan@unikl.edu.my

M. F. bin Abu Hassan
e-mail: mohdfauzi@unikl.edu.my

T. A. A. Razak et al. (eds.), *Progress in Engineering Technology VI*, Advanced Structured Materials 229, https://doi.org/10.1007/978-3-031-81517-1_30

30.1 Introduction

Hydrofluorocarbons (HFCs) have been part of fire suppression systems and are replacing halon where the issue of global warming now has an impact on climate change, thereby threatening global sustainability [1]. These agents extinguish the fire in total flooding [2] mode in an enclosed space. The role of its distribution nozzle is becoming an important factor to ensure the fire can be extinguished without disturbing the electrical system's operation and leaving no residue in the environment. Since HFC is considered hazardous, it is time to find another alternative agent with the same function that can be used [3]. Since the mechanical properties of HFCs and the proposed fire suppression agent are different, it affects the quality of turbulence flow of the agent towards the affected environment. Therefore, the effectiveness of the systems not only depends on the agent's liquid or vapour properties but also on the function of the distribution nozzle design, which contributes to the success of fire suppression. Hence, deciding the design of the distribution nozzle is important to reduce loss and increase its effectiveness.

Due to global warming issues of halocarbons, most researchers have discovered and nominated HFOs as hydrocarbon alternatives [1]. However, the application of distribution nozzle HFO's fire suppression for commercial use has not been studied yet. It is time to explore its function by adapting the practice conceptual design.

Theory and innovative problem solving (TRIZ) is a problem-solving tool that was designed to allow designers to find inventive and creative solutions [4, 5]. Many authors have documented the use of TRIZ for process improvement [6], and optimization and applied different techniques [7]. Even for the most complex problems, TRIZ provides faster and more innovative solutions. According to the findings, TRIZ approaches should be helpful for usage in practice and were recommended for this study [6]. As with other well-known products or prototypes, TRIZ has been widely proposed and explored to be used as a conceptual design [8–11]. However, it has not been applied to the development of a new fire suppression distribution nozzle prototype.

This study aims to determine the ideal solution or perfection for the development of distribution nozzle effectiveness. The TRIZ methodology is based on four fundamental concepts: (1) Contradictions (2) Perfection (3) Functionality (4) Making use of resources [12]. In this context, the use of TRIZ can solve the problem that occurs in the product or the problem that the material itself faces. For instance, the problem of global sustainability of hydrofluorocarbons can cause global warming issues [1], while the problem of the product, for example, the function of the distribution nozzle that distributes the gas towards the fire, can be improved. Therefore, TRIZ has been used to overcome this issue.

The application of TRIZ in the early stages assisted design engineers to simulate function through computational fluid modelling (CFD) before fabrication was made [13]. Integrating TRIZ and CFD should be a trend to solve mechanical design complexity [14] preventing redundancy, time loss, and mistakes once it becomes a real product.

This study uses TRIZ to design a distribution nozzle as a new hydrofluorocarbon fire suppression alternative, ensuring it is practically usable for prototype testing and eventually applicable as a product.

30.2 Application of TRIZ Method for the Development of the New Fire Suppression Distribution Nozzle

Conceptual design is a strategic approach to revolutionizing industries by using technology or products. It involves four fundamental phases, as illustrated in Fig. 30.1. Understanding the context, target market and product, outlining ideas in shapes, usability, and prototypes. During the refining phase, it is crucial to incorporate design language and reality to effectively present the design. Designer ensure files are thoroughly delivered and translated to technical workgroups and manufacturers to minimize perceptions of intention, crucial for manufacturing processes [15].

Then, the detailing of the engineering parameters was refined using the TRIZ contradiction matrix to provide inventive solutions for the product [5]. A final decision can be made by transferring the concepts highlighted in the matrix to computer-aided design and CFD works to simulate fluid flow.

Initially, there are two primary options for solving problems: either enhancing the functionality of the distribution nozzle or improving the properties of the gas used as a fire suppression agent. Both options are crucial considerations in the decision to apply the TRIZ method. Enhancing the nozzle's functionality could lead to more efficient and targeted distribution of the suppression agent, increasing the effectiveness of fire suppression efforts. On the other hand, improving the properties of the gas itself could result in a more potent and reliable agent, capable of extinguishing fires more rapidly and effectively. Given their significant impact, both options warrant thorough evaluation and analysis within the TRIZ framework to determine the most effective solution.

To analyse and suggest which principles should be used for the development of the distribution nozzle, the problem statement must be clearly defined, as shown in Fig. 30.2. A statement of conflict was made by assigning "what to improve" to justify what parameter was to be improved and one worsening parameter involved in the matrix mapping. The result shows the chosen principles as guidance for the development of the prototype.

Based on the decision made, a relevant problem statement was formulated and a statement of conflict has been identified where the improvement of the vaporized

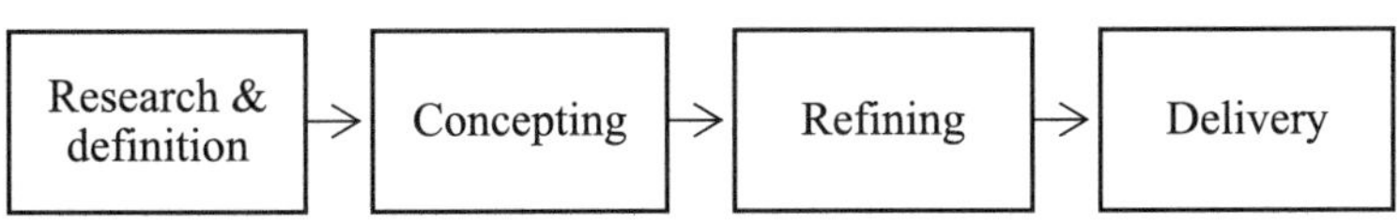

Fig. 30.1 Phases in conceptual design

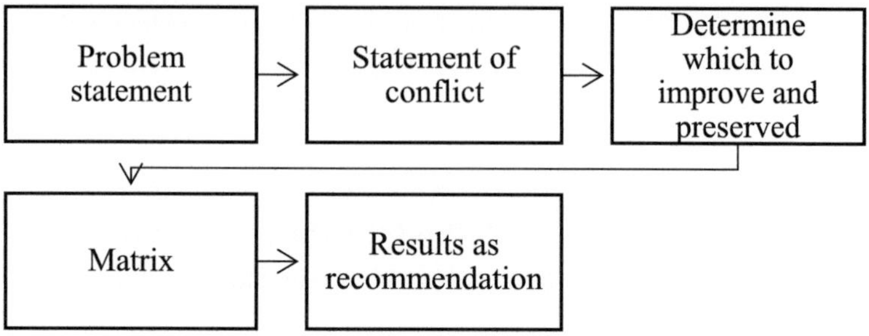

Fig. 30.2 TRIZ flow for the development, recommendation for a new distribution nozzle

agent has been thoroughly distributed in the cubical but is not affecting its standard time. Then, the features to improve and the features to preserve were selected based on 39 features.

The flowchart in Fig. 30.3 illustrates the TRIZ implementation process in the design phase. It begins with defining the problem and identifying features to improve and preserve. Solutions are generated using the TRIZ matrix, followed by verification through Computational Fluid Dynamics (CFD) simulations. Finally, the findings are evaluated to ensure the solutions effectively address the problem and meet the design goals.

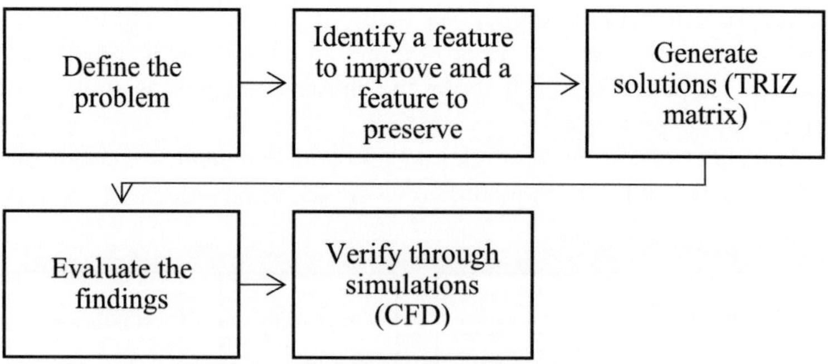

Fig. 30.3 The flow of TRIZ implementation until the design has been simulated to ensure the design met its purpose

30.3 Results

This section exposes the results of the five-step process, revealing 39 combinations of improved and preserved matrix feature options that overcame contradictions. These included nine common inventive principles that matched the design's requirements. They are then filtered and discovered to have overcome 39 contradictions, almost matching the criteria. Finally, the contradiction was resolved by a pair of improved and preserved matrix features comprised of three inventive principles, as shown in Table 30.1.

Table 30.2 represents the mapping of improving energy loss, which affects the speed of fluid flow out of the nozzle outlet as well as the turbulent flow. To address this issue, a series of parameter settings are proposed to improve its efficiency using CFD modelling.

The study proposed three strategies for developing a prototype, as shown in Table 30.3 incorporating partial or excessive action to optimize functionality, modifying parameters for efficiency and adaptability and ensuring robustness through a strong oxidant. These strategies aim to drive innovation and deliver a high-quality, reliable product to meet customer needs.

Partial or excessive action in fire extinguishment optimized gas distribution of halocarbon alternatives improved suppression outcomes. This approach covers a large area, customizes for various fire types and sizes. This reasonable approach enhances safety, resource utilization, and overall effectiveness in fire extinguishment techniques.

To maintain constant fluid turbulence in fire suppression systems, parameter changes, such as modifying orifices or outlets, can be justified. This ensures optimal coverage, improves extinguishing efficiency, and allows the system to operate in specific environments, leading to improved fire suppression performance and enhanced safety.

Strong oxidants propose the right choice of material to ensure the proposed distribution nozzle can work after frequent usage. This eases maintenance and prevents corrosion inside the nozzle, which affects its effectiveness during use.

Table 30.1 Set of contradictions to solve for the HFC alternative fire suppression distribution nozzle focusing on the needs of its application

	Feature to improve	Feature to preserve
Features	Loss of energy	Speed
Rational	Improving the energy of the fluid instantly reduces the fire's temperature	Maintain the timing of extinguishment based on the standards

Table 30.2 Summary of TRIZ matrix mapping for the development of distribution nozzle hydrocarbon alternative

	Features	PRESERVE							
		7: Volume of moving object	8: Volume of stationary	**9: Speed**	10: Force (Intensity)	11: Stress of pressure	12: Shape	13: Stability of object	14: Strength
IMPROVING	9: Speed	7 29 34	–		13 28 15 19	6 18 38 40	35 15 18 34	28 33 1 18	8 3 26 14
	10: Force (intensity)	15 9 12 37	2 36 18 37	13 28 15 12		18 21 11	10 35 40 34	35 10 21	35 10 14 27
	11: Stress of pressure	6 35 10	35 24	6 35 36	35 36 21		35 4 15 10	35 33 2 40	9 18 3 40
	12: Shape	14 4 15 22	7 2 35	35 15 34 18	35 10 37 40	34 15 10 14		33 1 18 4	30 14 10 40
	13: Stability of object	28 10 19 39	34 28 35 40	33 15 28 18	10 35 21 16	2 35 40	22 1 18 4		17 9 15
	14: Strength	10 15 14 7	9 14 17 15	8 13 26 14	10 18 3 14	10 3 18 40	10 30 35 40	13 17 35	
	17: Temperature	34 39 40 18	35 6 4	2 28 36 30	35 10 3 21	35 39 19 2	14 22 19 32	1 35 32	10 30 22 40
	21: Power	35 6 38	30 6 25	15 35 2	26 2 36 35	22 10 35	29 14 2 40	35 32 15 31	26 10 28
	22: Loss of energy	7 18 23	7	**16 35 38**	36 38	–	–	14 2 39 6	26

The content in bold indicates the mapped features for improvement and preservation for the dedicated design. The numbers show the necessary principles required for designated design.

Table 30.3 The TRIZ matrix proposes the following principles to solve this contradiction

Principle No.	Principle	Explanation
16	Partial or excessive action	• 100% of an object is difficult to obtain using a given solution method, use "slightly more or less" of the same method, the problem may be easier to solve
35	Parameter changes	• Change the physical state of fluid • Change fluid concentration/density • Change the temperature/volume • Change the degree of flexibility
38	Strong oxidants	• Fabricating selected alloy nozzle

30.4 Conclusion

A distribution nozzle prototype has been developed and built based on the proposed principle. The study proposed a solution for controlling the distribution of vaporized agents while extinguishing a fire in an enclosed environment.

Based on the contradiction matrix analysis, three innovative principles were proposed to address the identified challenges: partial or excessive action, parameter changes, and the use of a strong oxidant. Among these, the principle of parameter change has been emphasized as the most critical for advancing the development of a new distribution nozzle designed for hydrofluorocarbon fire suppression alternatives. This focus on parameter changes aims to optimize the nozzle's performance and efficiency, which is essential for improving the effectiveness of fire suppression systems.

The suggested distribution nozzle is made of brass or stainless steel and consists of multiple outlets with appropriate angles. The design fulfils TRIZ principles. The outlet diameter, degree of flexibility, and material were designed to ensure its effectiveness in extinguishing fire. These will be incorporated into computational modelling to validate its function and increase its effectiveness.

The study proposed that the chosen concept should be used during design, simulation for prototypes development and fabrication of a new distribution nozzle as a hydrocarbon alternative, which is expected to be simulated using CFD and patented before being commercialized.

References

1. Chimote RS, Singh V, Yadav S, Yadava R (2018) Impact of HFC fire extinguishing clean agents on climate change and its system design requirements for fire hazards in India—A brief study. Clim Change Impacts Water Sci Technol Lib 82:293–317
2. Forssell EW, Hill S, Kramer B, Selecting the correct clean agent suppression system. CSEMAG. https://www.csemag.com/articles/selecting-the-

3. Heredia-Aricapa Y, Belman-Flores JM, Mota-Babiloni A, Serrano-Arellano J, García-Pabón JJ (2020) Overview of low GWP mixtures for the replacement of HFC refrigerants: R134a, R404A and R410A. IJR, pp 113–123
4. Sheu DD, Chiu MC, Cayard D (2020) The 7 pillars of TRIZ philosophies. Comput Ind Eng 146
5. Asyraf MRM et al (2020) Integration of TRIZ, morphological chart and ANP method for development of FRP composite portable fire extinguisher. Polym Compos 41:2917–2932
6. Sojka V, Lepsik P (2020) Use of TRIZ, and TRIZ with other tools for process improvement: a literature review. Emerg Sci J
7. Asyraf MRM, Ishak MR, Sapuan SM, Yidris N (2019) Conceptual design of creep testing rig for full-scale cross arm using TRIZ-Morphological chart-analytic network process technique. J Mater Res Technol 8(6):5647–5658
8. Asyraf MRM, Ishak MR, Sapuan SM, Yidris N (2020) Conceptual design of multi-operation outdoor flexural creep test rig using hybrid concurrent engineering approach. J Mater Res Technol 9(2):2357–2368
9. Sharaf HK, Ishak MR, Sapuan SM, Yidris N (2020) Conceptual design of the cross-arm for the application in the transmission towers by using TRIZ–morphological chart–ANP methods. J Mater Res Technol 9(4):9182–9188
10. Fiorineschi L, Frillici FS, Rotini F (2018) Supporting systematic conceptual design with triz. Proc Int Design Conf 3:1091–1102
11. Zhang J, Chai KH, Tan KC (2005) Applying TRIZ to service conceptual design: an exploratory study. Creat Innov Manag 14(1)
12. Ekmekci I, Nebati EE (2019) Triz methodology and applications. Proc Comp Sci 158:303–315
13. Uzoka C, Mishra R (2020) Integration of TRIZ and CFD to new product development process. Int J Comput Fluid Dyn 34(6):418–437
14. Fiorineschi L, Frillici FS, Rotini F (2018) Enhancing functional decomposition and morphology with TRIZ: Literature review. Comput Ind 94:1–15
15. Amir AL, Ishak MR, Yidris N, Zuhri MYM, Asyraf MRM (2021) Advances of composite cross arms with incorporation of material core structures: manufacturability, recent progress and views. J Mater Res Technol 13:1115–1131

Chapter 31
Optimization of a Car Rear Spoiler Using Bezier Curves

Mohammad Syakir Fathullah Bin Shuib, Mazwin Binti Tan, Johan Ihsan Mahmood, John Jabaraj Devadason, and Mohd Fauzi Abu Hassan

Abstract This project investigates the optimization of the aerodynamic design of a car rear spoiler. Several types of designs have been chosen in this project to make a comparison. A Bezier curve fitting approach is used as a tool to improve the existing design of the rear spoiler. A formal optimization technique based on computational fluid dynamics (CFD) is used to determine the control points for the approximate curve that covers the rear spoiler. It is hoped that with this analysis, the designation of the rear spoiler will be one of the factors to be considered by car users to choose their rear spoiler to reduce drag and hence improve the car's performance.

Keywords Computational fluid dynamic (CFD) · Bezier curve · Airfoil · Shape approximation

31.1 Introduction

A Bezier curve can be described using a mathematical formulation. It is a parametric curve defined by a set of control points, two of which are the ends of the curve, and the others determine its shape. Given the coordinates of control points P_i: the first

M. S. F. B. Shuib · M. B. Tan (✉) · J. I. Mahmood · J. J. Devadason · M. F. A. Hassan
Kampus Cawangan Malaysian Spanish Institute Kulim Hi Tech Park, Universiti Kuala Lumpur, 09000 Kulim, Kedah, Malaysia
e-mail: mazwin@unikl.edu.my

M. S. F. B. Shuib
e-mail: syakir.shuib@s.unikl.edu.my

J. I. Mahmood
e-mail: johanihsan@unikl.edu.my

J. J. Devadason
e-mail: john@unikl.edu.my

M. F. A. Hassan
e-mail: mohdfauzi@unikl.edu.my

T. A. A. Razak et al. (eds.), *Progress in Engineering Technology VI*, Advanced Structured Materials 229, https://doi.org/10.1007/978-3-031-81517-1_31

control point has coordinates $P_1 = (x_1, y_1)$, the second: $P_2 = (x_2, y_2)$, and so on, the curve coordinates are described by the equation that depends on the parameter t from the segment [0,1]. In this paper, the four control points formula has been used to ensure the smoothness of the design:

$$P = (1 - t)^3 P_1 + 3(1 - t)^2 P_2 + 3(1 - t)t^2 P_3 + t^3 P_4$$

Bezier curves are a powerful mechanism for modeling in computer-aided geometric design (CAGD) and computer graphics (CG). Bezier curves have a lot of applications in the areas of science, engineering, and technology such as railway route or highway modeling, networks, animation, computer-aided design systems, robotics, environment design, communications, and many other fields due to their computational simplicity and stability. However, the classical Bezier curves still have some noiseless limitations due to their fixed shape and position relative to their control polygon. In geometric modeling and engineering, practical applications of Bezier curves are restricted due to their shortcomings, and to overcome these shortcomings, a lot of work has been carried out. The control over the shape and position of curves is enhanced by introducing the shape control parameters into the Bezier approach. Another limitation of the classical Bezier curves is their representation in a polynomial form.

31.2 Literature

In the 1970s, a mechanical engineer from Renault Car Company created the Bezier representation. The basic idea is to write the parametric polynomial in such a way that the shape control parameters are described clearly by control points that do not have to be on the curve. The manual approach is used to approximate the shape of an airfoil using a cubic Bezier curve and a quintic Bezier curve. The cubic and quintic Bezier curve approximations can both satisfy the 0.1 mm tolerance. However, the degree of freedom of a cubic Bezier curve is inadequate to describe the shape of a highly curving region. The shape of a highly curving region can be represented using a quintic Bezier curve, which enables second-derivative control [1]. The findings and discussions in this literature review are supported by the cited works [2–6], ensuring a comprehensive understanding of Bézier representations and their applications.

31.2.1 Airfoil Shape Approximation by a Cubic Bezier Curve

This subsection identifies the cubic Bezier curve's approximation results for airfoils. The control points that result are given by $V_0(1.9, 1)$, $V_1(0.8, 1)$, $V_2(-0.64, 4)$, and $V_3(6.1, 1)$ as shown in Fig. 31.1.

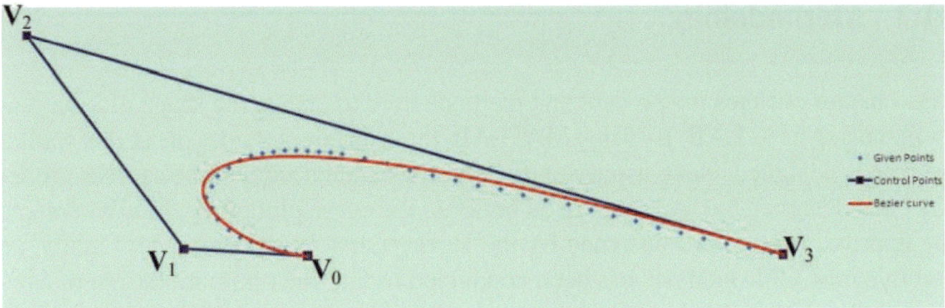

Fig. 31.1 Cubic Bezier curve that is approximating an airfoil shape

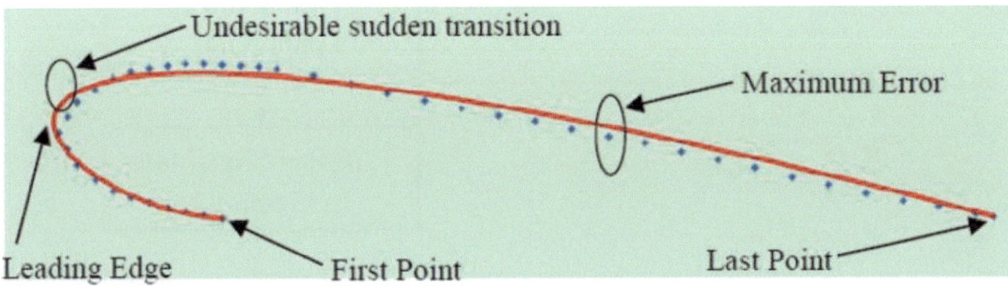

Fig. 31.2 Enlargement view of the approximating airfoil shape

As shown in Fig. 31.2, the largest error of 0.093 mm occurs in the flatter section of the airfoil, which falls within the allowable tolerance of 0.1 mm. The most difficult section of the approximation is at the leading edge of the airfoil form, which is mostly governed by control points V_1 and V_2. The control point V_1 is concerned with the tangent of the first point, which defines the curve's bending direction.

31.2.2 Problem

The Bezier representation is a parametric function that is commonly used to approximate free-form shapes. It is preferable to approximate the shape with Bezier representation given a string of points with the assumption that they are sufficiently dense to characterize the shape of an airfoil. Within an acceptable operating tolerance, the representation function should be near the original shape design [1]. By using the Bezier curve as a tool to shape design, constructing full-scale drawings was no longer needed.

31.3 Methodology

This chapter outlines the process and methods used to design the shape of a car rear spoiler using MATLAB. By using MATLAB, three different shapes of car rear spoiler have been designed. Next, using SolidWorks, the actual design of the car rear spoiler has been designed too. A comparison between the curve plotted by SolidWorks and the approximated curve designed by the Bezier curve has been made. Finally, an aerodynamic CFD analysis has been conducted to find the optimum design of car's rear spoiler. Figure 31.3 shows the process of the development of this study.

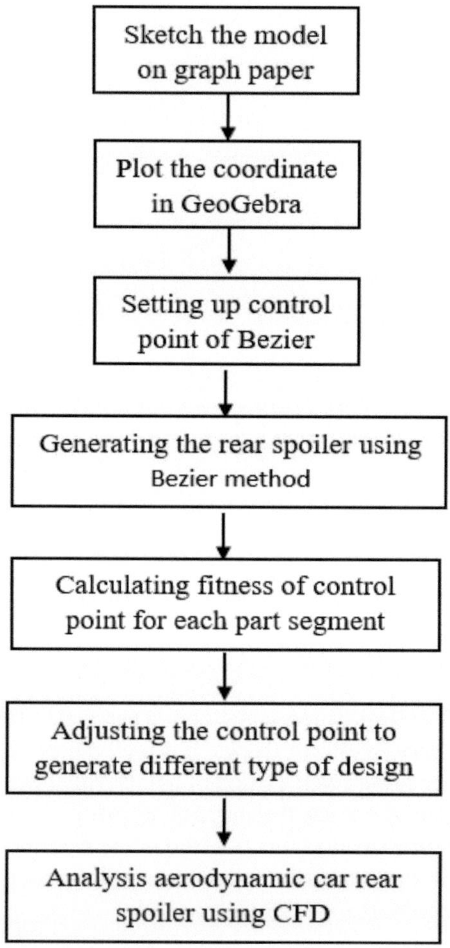

Fig. 31.3 Flowchart of project development

31.3.1 Research Method

In this paper, GeoGebra was used to determine the control points of the rear spoiler (airfoil) to generate Bezier curve control points in MATLAB. This is the first step in shaping the airfoil for the car's rear spoiler. Meshing is one of the key components to obtaining accurate results from a numerical model. The computational fluid dynamics (CFD) method is used in this project to optimize each airfoil design. To obtain the results, meshing techniques and boundary conditions were applied in this section. All the airfoil meshing designs, as shown in Figs. 31.4, 31.5 and 31.6, were created using the same conditions. The inlet velocity and outflow pressure were set at line 4 and line 2, respectively, while the grid size for the meshing was set to 0.4

The inputs to the simulation model are defined by the boundary conditions. These conditions, which determine how a fluid enters or exits the model, include parameters like velocity and volumetric flow rate. The exchange of energy between the model and its surroundings is defined by factors such as the film coefficient and heat flow.

31.3.2 Optimization Method

The objective is to minimize the aerodynamic drag force D (in Newtons) given by $\mathrm{Cd} = \frac{2\mathrm{Fd}}{\rho V^2 A}$, where ρ is the air density, A is the frontal area, V is the free-stream velocity, and Cd is the drag coefficient. Since D, for a given density and velocity, is proportional to the product of the vehicle's frontal area and the drag coefficient, this objective function represents a compromise between the size and aerodynamic performance of the car

In this project, a BMW M4 series car model was chosen due to its performance. This model was designed in SolidWorks in three dimensions to enable a full aerodynamic simulation. The analysis was conducted using a rear spoiler with three different shapes and without a rear spoiler to observe the different values of drag force and drag coefficient. Table 31.1 shows the details of the values that had been used in this study with three different frontal area values to calculate the drag force.

31.3.3 Simulation Model with MATLAB/Simulink

In this project, CFD simulations use the Navier–Stokes equations, which are implemented in the computational fluid dynamics FEA Tool Multiphysics software.

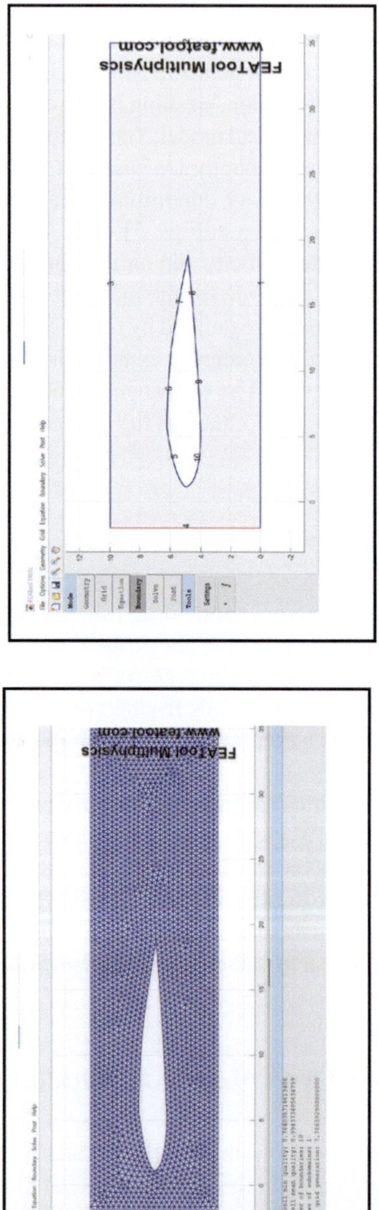

Fig. 31.4 Meshing technique and boundary condition of airfoil shape design 1

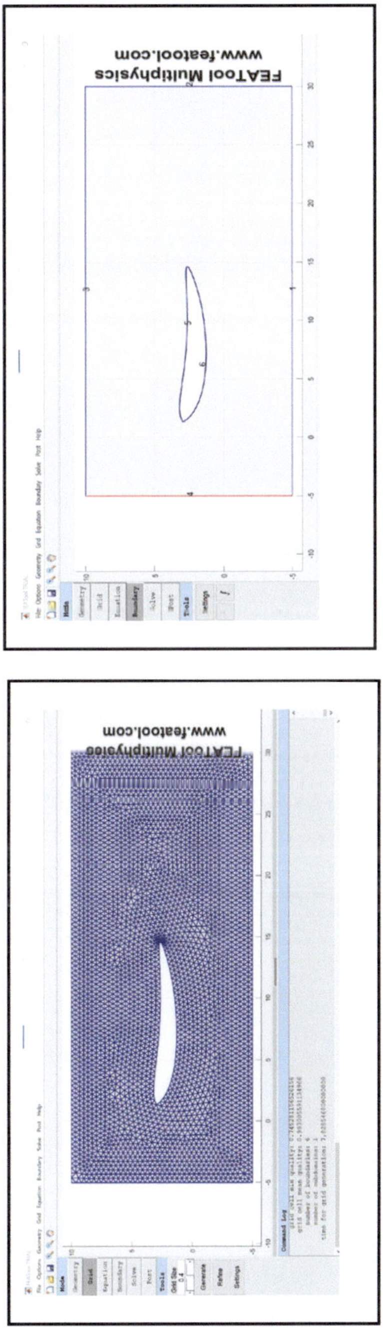

Fig. 31.5 Meshing technique and boundary condition of airfoil shape design 2

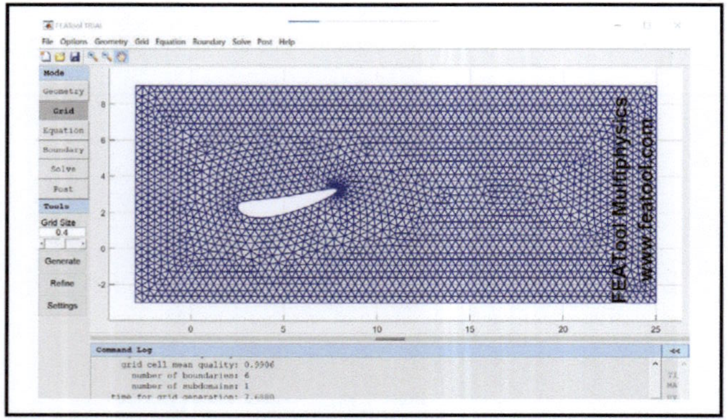

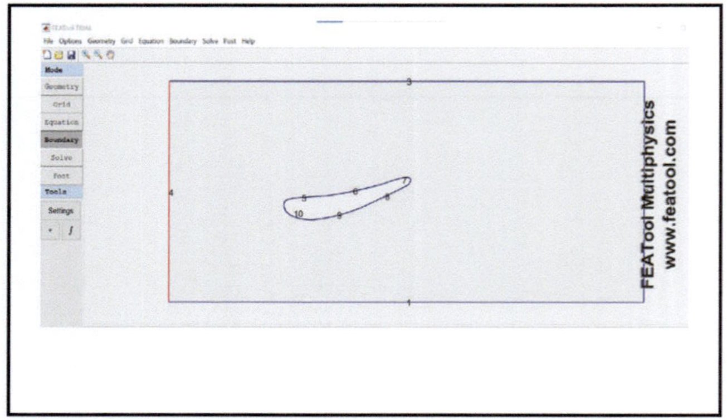

Fig. 31.6 Meshing technique and boundary condition of airfoil shape design 3

Table 31.1 The values of the parameters that had been used to calculate the drag force, F_d

F_d	GG Force Z
ρ	1.2 kg/m^3
V	55.556 m/s
A	Base car—8 m^3 Airfoil design 1—8.17 m^2 Airfoil design 2—8.5 m^2 Airfoil design 3—10 m^2

31.4 Results and Discussion

Figures 31.7 and 31.8 show the drag force for airfoil design 1, which was plotted using the Bezier curve and SolidWorks, respectively. The values of drag force for both designs are shown in Table 31.2 for comparison. From this table, it can be concluded that the drag force value plotted using the Bezier curve was better than the value plotted using SolidWorks.

31.4.1 Optimization Result

Based on Figs. 31.10, 31.11, 31.12 and 31.13, it is shown that cars with no spoiler have a higher value of drag force (*N*) than model cars that have spoiler. The reduction in drag for road cars can enhance handling, acceleration, and many other performance characteristics, as well as the vehicle's top speed and fuel efficiency. The frontal area of the vehicle and the drag coefficient are the two fundamental parameters that affect the drag.

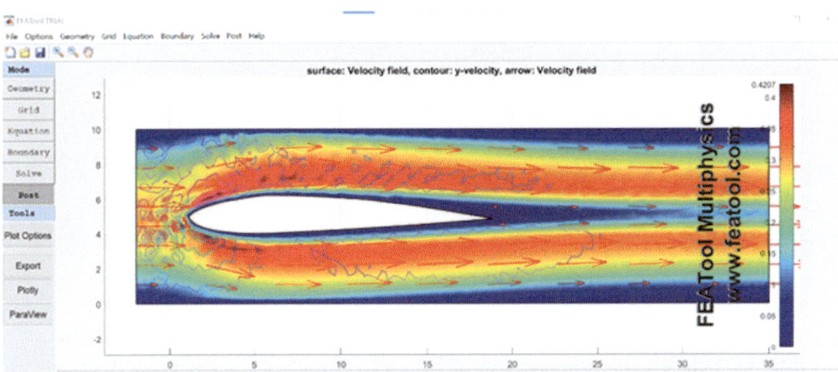

Fig. 31.7 Drag force for airfoil design 1 plotted by Bezier curve

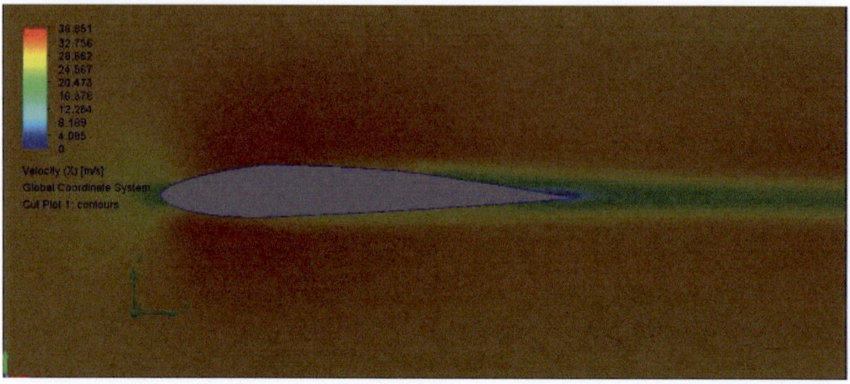

Fig. 31.8 Airfoil design 1 plotted by actual curve

Table 31.2 Comparison of drag force value of the design plotted by the Bezier and the SolidWorks

Curve	Drag force (*N*)
Bezier curve (Fig. 31.7)	32.1416
Actual curve (Fig. 31.8)	36.851

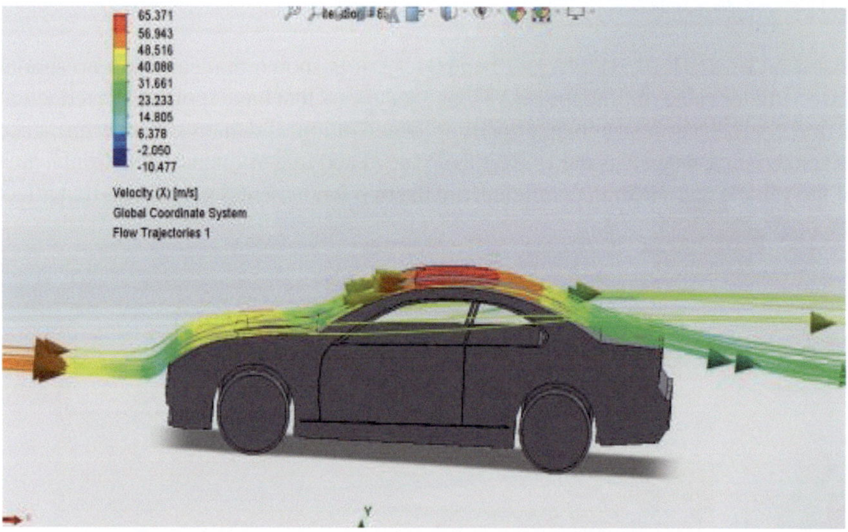

Fig. 31.10 BMW model with no spoiler

Table 31.3 shows the comparison of the drag force values between the basic car (car without a spoiler) and the three different designs of the cars with spoilers. The airflows were simulated at a constant value of 55.556 m/s. From the analysis, it can be concluded that having a rear spoiler improves the car's efficiency and performance.

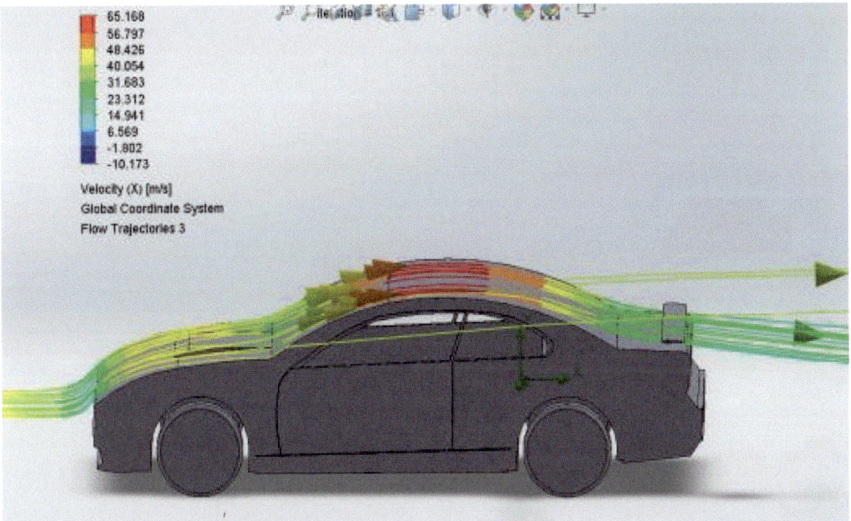

Fig. 31.11 BMW model with airfoil design 1

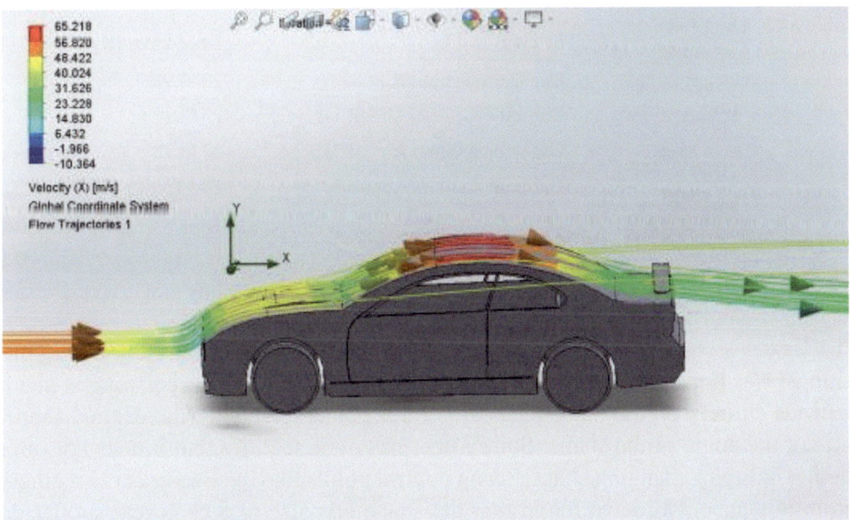

Fig. 31.12 BMW model with airfoil design 2

Additionally, Table 31.3 indicates that among the three designs, airfoil design 1 has the lowest drag force, which potentially improves car performance and reduces fuel consumption.

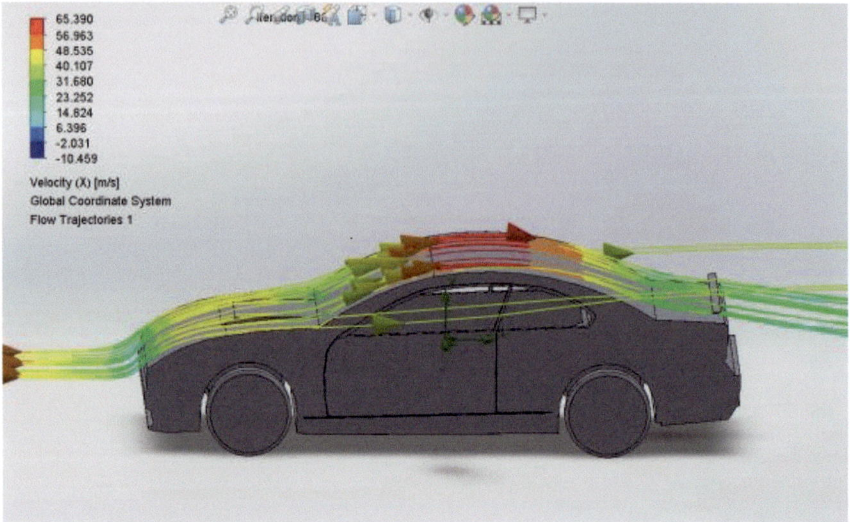

Fig. 31.13 BMW model with airfoil design 3

Table 31.3 Drag force comparison for cars without and with spoilers of three different designs

Design	Drag force (N)
Car without spoiler	2349.93
Car with spoiler (Airfoil 1)	2343.13
Car with spoiler (Airfoil 2)	2348.78
Car with spoiler (Airfoil 3)	2349.44

31.5 Conclusions

The Bezier technique, including Bezier curves, their theory, properties, and applications, was implemented using MATLAB. The limitation of this software is that it depends entirely on the adjustment of control points to achieve the desired shape, but not the angle of the shape. SolidWorks has more features, but it does not offer the flexibility in changing shapes using control points like the Bezier curve method. From this paper, we have found that the shape and size of a car's rear spoiler do influence the car's performance. Therefore, it is crucial to study the design of each part of the car that has the potential to reduce drag force and thus improve the car's efficiency.

References

1. Lin C, Perk YC (2013) Approximation of free-form curve-airfoil shape. J Eng Sci Technol 8(6)

2. Taherkhani AR, Gilkeson C, Gaskell P, Hewson R, Toropov V, Rezaienia A, Thompson H (2017) Aerodynamic CFD-based optimization of police car using Bezier curves. SAE Int J Mater Manuf 10(2):85–93. https://doi.org/10.4271/2017-01-1548
3. Li H, Luo Y, Wu J (2019) Collision-free path planning for intelligent vehicles based on Bézier curve. IEEE Access 7:123334–123340. https://doi.org/10.1109/ACCESS.2019.2938654
4. Yomchinda T (2018) Simplified propeller model for the study of UAV aerodynamics using CFD method. In: 5th Asian conference on defense technology (ACDT), pp 69–74. https://doi.org/10.1109/ACDT.2018.8394321
5. Miguel VE, Victor JAGG, Omar MJ, Pedro P, Arturo M (2021) Deep Q-learning for control: technique and implementation considerations on a physical system—active automotive rear spoiler case. In: 2021 IEEE International conference on mechatronics and automation (ICMA), pp 818–824. https://doi.org/10.1109/ICMA52036.2021.9512713
6. Eftekhari H, Al-Obaidi ASM, Eftekhari S (2020) The effect of spoiler shape and setting angle on racing cars' aerodynamic performance. Int J Sci Technol 5(1):11–20

MIX
Papier aus verantwortungsvollen Quellen
Paper from responsible sources
FSC® C105338

If you have any concerns about our products,
you can contact us on
ProductSafety@springernature.com

In case Publisher is established outside the EU,
the EU authorized representative is:
Springer Nature Customer Service Center GmbH
Europaplatz 3, 69115 Heidelberg, Germany

Printed by Libri Plureos GmbH
in Hamburg, Germany